SAFETY AND RISK IN SOCIETY

PILOT TRAINING AND SAFETY

CONSIDERATIONS AND ASSESSMENTS

SAFETY AND RISK IN SOCIETY

Additional books in this series can be found on Nova's website under the Series tab.

Additional E-books in this series can be found on Nova's website under the E-book tab.

SAFETY AND RISK IN SOCIETY

PILOT TRAINING AND SAFETY

CONSIDERATIONS AND ASSESSMENTS

DAVID M. FONTENELLE
EDITOR

NOVINKA

New York

For permission to use material from this book please contact us:
Telephone 631-231-7269; Fax 631-231-8175
Web Site: http://www.novapublishers.com

Library of Congress Cataloging-in-Publication Data

ISBN: 978-1-62417-400-1

Published by Nova Science Publishers, Inc. † New York

CONTENTS

PREFACE

The United States' aviation transportation system is an important component of the U.S. economy and one of the safest and most efficient systems in the world. However, the continued integrity of this system depends in part on the roughly 3,400 U.S. pilot schools providing well-trained pilot candidates for airlines. This is particularly true for regional airlines, which hire and train pilots from these schools, whereas large mainline airlines tend to hire and train pilots who already have experience at regional airlines. Although the large mainline airlines have an unprecedented safety record, in recent years regional airlines, which account for about 53% of the nation's commercial flights, have experienced a number of fatal accidents; specifically the last six fatal commercial airline accidents involved regional airlines. This book discusses the various types of U.S. pilot training organizations, how they compare, and issues associated with training. Also discussed is to what extent the FAA carries out its oversight role of pilot training and certification.

Chapter 1 – Regional airlines have experienced the last six fatal commercial airline accidents, and pilot performance has been cited as a potential contributory factor in four of these accidents. As a result, Congress and others have raised questions about, among other issues, the initial pilot education and training required before pilots can be hired by airlines, at which time they receive further training. The initial training is provided by pilot schools overseen by the Federal Aviation Administration (FAA).

As requested, this report discusses (1) the various types of U.S. pilot schools, how they compare, and associated issues; (2) key similarities and differences between the U.S. and international approaches to pilot training; and (3) how and to what extent FAA carries out oversight of pilot training and certification. To address these issues, GAO reviewed literature, legislation, regulations, and FAA documents and inspection and enforcement data;

interviewed agency and industry officials; and studied the training approach in Europe because of the different training model and visited four European countries.

Chapter 2 – FAA has met or is on schedule to meet many of the Act's requirements, such as improving pilot rest requirements and establishing better processes for managing safety risks.

However, FAA has not met timelines for raising pilot training standards, implementing mentoring programs, providing enhanced leadership skills to captains, and increasing minimum pilot qualifications. FAA also faces challenges in establishing a pilot records database—an important component for enhancing the air carrier screening process for pilot applicants. In addition to overcoming these challenges, FAA needs to provide additional guidance and assistance to industry— especially smaller carriers—in developing and managing new safety programs.

Chapter 3 – Testimony of William R. Voss, President and CEO, Flight Safety Foundation.

Chapter 4 – Statement of Captain Carl Kuwitzky, President, Coalition of Airline Pilots Associations

In: Pilot Training and Safety
Editor: David M. Fontenelle
ISBN: 978-1-62417-400-1

Chapter 1

INITIAL PILOT TRAINING: BETTER MANAGEMENT CONTROLS ARE NEEDED TO IMPROVE FAA OVERSIGHT*

United States Government Accountability Office

ABBREVIATIONS

AABI	Aviation Accreditation Board International
ALPA	Air Line Pilots Association
AOPA	Aircraft Owners and Pilots Association
CAPA	Coalition of Airline Pilots Associations
EASA	European Aviation Safety Agency
EIS	Enforcement Information System
FAA	Federal Aviation Administration
FITS	FAA Industry Training Standards
IATA	International Air Transport Association
NASA	National Aeronautics and Space Administration
NPG	National Program Guidelines
NTSB	National Transportation Safety Board
PTRS	Program Tracking and Reporting Subsystem

* This is an edited, reformatted and augmented version of The United States Government Accountability Office publication, Report to Congressional Requesters GAO-12-117, dated November 2011.

SPAS	Safety Performance Analysis System
UAA	University Aviation Association

Why GAO Did This Study

Regional airlines have experienced the last six fatal commercial airline accidents, and pilot performance has been cited as a potential contributory factor in four of these accidents. As a result, Congress and others have raised questions about, among other issues, the initial pilot education and training required before pilots can be hired by airlines, at which time they receive further training. The initial training is provided by pilot schools overseen by the Federal Aviation Administration (FAA).

As requested, this report discusses (1) the various types of U.S. pilot schools, how they compare, and associated issues; (2) key similarities and differences between the U.S. and international approaches to pilot training; and (3) how and to what extent FAA carries out oversight of pilot training and certification. To address these issues, GAO reviewed literature, legislation, regulations, and FAA documents and inspection and enforcement data; interviewed agency and industry officials; and studied the training approach in Europe because of the different training model and visited four European countries.

What GAO Recommends

To improve oversight of pilot certification, GAO recommends that FAA develop a comprehensive system to (1) measure its performance in meeting the agency's annual inspection requirements for pilot schools and pilot examiners and (2) better understanding the nature and scope of discretionary inspections for flight instructors. FAA generally agreed with our recommendations.

What GAO Found

The approximately 3,400 pilot schools in the United States can be divided into three types: (1) flight instructor based, (2) vocational, and (3) collegiate.

The school types vary in several ways, but all pilot students must pass the same knowledge and flight tests to obtain a pilot certificate from FAA. Airline operations have evolved operationally and technologically, but the pilot training requirements for certification of commercial pilots were last revised in 1997. FAA and some industry stakeholders have indicated that current requirements for commercial pilots should incorporate additional training to improve the competency of entry-level regional airline pilots. FAA has initiated or planned a number of efforts to address these issues and recently enacted legislation requires FAA to implement regulations to increase pilot requirements for airlines by August 2013.

Sources: Cessna (single-engine trainer), Piper (multi-engine trainer), and Bombadier (regional jet).

Example of a Single- and Multi-Engine Training and a Regional Airline Jet.

The U.S. and Europe both offer the same pilot certifications but the training models differ, in part, due to training philosophies and other circumstances. The U.S. training approach emphasizes proficiency on actual flight training, while Europe's approach tends to emphasize academic instruction with more knowledge training requirements and testing. European pilot schools have also developed more comprehensive student screening processes than in the U.S.

FAA has an annual inspection program that includes the oversight of pilot schools, pilot examiners, and flight instructors, the gatekeepers for the initial pilot training process. GAO analysis of FAA inspection data showed a 78 percent completion rate of the required inspections for pilot schools in fiscal year 2010, but, due to insufficient information, GAO was unable to determine completion percentages for prior years. Similarly, GAO could not determine 1) whether FAA completed the required inspections for pilot examiners or 2) the reasons that the discretionary inspections of flight instructors—which are generally optional—were conducted. Furthermore, FAA's national office does not adequately monitor the completion of annual inspection activities due, in part, to an inability to aggregate inspection data from the local district offices that conduct the inspections. Thus, FAA does not have a comprehensive system in place to adequately measure its performance in meeting annual inspection requirements, which could make it difficult to ensure regulatory compliance and that safety standards are being met.

November 4, 2011

Congressional Requesters

The United States' aviation transportation system is an important component of the U.S. economy and one of the safest and most efficient systems in the world. However, the continued integrity of this system depends in part on the roughly 3,400 U.S. pilot schools providing well-trained pilot candidates for airlines. This is particularly true for regional airlines, which hire and train pilots from these schools, whereas large mainline airlines tend to hire and train pilots who already have experience at regional airlines.[1] Although the large mainline airlines have an unprecedented safety record, in recent years regional airlines, which account for about 53 percent of the nation's commercial flights,[2] have experienced a number of fatal accidents; specifically, the last six fatal commercial airline accidents involved regional airlines. The National Transportation Safety Board (NTSB) has cited pilot performance as a potential contributory factor in four of these accidents, including the one in February 2009, in Buffalo, New York, involving Colgan Air, Inc.[3] As a result, Congress and others have raised questions about, among other issues, the initial pilot education and training required before pilots can be hired by airlines, at which time they receive further training. The initial training is provided by pilot schools overseen by the Federal Aviation Administration (FAA).[4]

You asked us to examine the initial pilot training system in the United States, which includes private and commercial pilot certification (license), up to the training provided once a pilot is hired by an airline. Because other countries have approached initial training for future airline pilots differently, we also studied the training approach in Europe to help inform our review. This report discusses (1) the various types of U.S. pilot training organizations, how they compare, and issues associated with training; (2) the key similarities and differences between the U.S. system and international approaches to pilot training; and (3) how and to what extent FAA carries out its oversight role of pilot training and certification. In addition, you noted that the worldwide demand for commercial airline pilots has also caused concerns to be raised about the ability of the current training system to provide an adequate number of pilots and asked that we examine this issue. We examined what is known about the future demand for and supply of U.S. commercial airline pilots and report our findings on this issue in appendix II.

To address our objectives and the pilot supply appendix, we reviewed and synthesized published literature related to pilot certification and training issues in the United States and comparable international pilot training systems; pilot statistics; types and requirements of pilot training schools; FAA regulatory training requirements for different levels of pilot certification; FAA oversight of U.S. pilot training system. Literature sources included federal agency documentation, academic journals, trade publications, conference presentations, and newspaper articles. We also reviewed the Federal Aviation Regulations related to training and certification for pilots, and legislative provisions that addressed issues related to pilot training. Specifically, we reviewed the Airline Safety and Federal Aviation Administration Extension Act of 2010 related to pilot training requirements and the certification of pilots operating for airlines.[5] We interviewed officials at the Department of Transportation (DOT), FAA, and NTSB. We conducted semistructured interviews with representatives from a range of aviation stakeholder organizations, including pilot unions, pilot school associations, general aviation groups, commercial aviation industry associations, international aviation associations, and regional airlines (see app. I for the complete list of stakeholders we interviewed). We visited pilot training stakeholders in six states—Arizona, Florida, Georgia, Indiana, Maryland, and Utah—which included FAA regions that had higher number of pilot schools, higher number of pilot certificates issued in 2009, presence of FAA regional and district offices, and presence of regional airlines' offices. In our state visits, we interviewed officials at FAA regional and district offices, regional airlines, pilot schools, and college aviation schools. Because we selected these six states as part of a nonprobability sample, our site visit data cannot be generalized to the nation. Through the combination of site visits and semistructured telephone interviews, we interviewed representatives of 24 regional airlines that transported about 97 percent of regional passengers in 2009, according to the Regional Airlines Association's 2010 annual report. We also analyzed data from FAA's data systems for fiscal years 2006 through 2010 that provided information on the oversight activities related to pilot certification and training. We tested the reliability of the inspection and enforcement data that we received from FAA with electronic testing of the data elements that we used, obtained, and reviewed documentation about the data and the systems that produced them, and interviewed knowledgeable FAA officials. We found the data to be sufficiently reliable for our purposes.

In addition, we conducted site visits to a nonprobability sample of European countries: France, Germany, the Netherlands, and the United Kingdom. We focused on the European pilot certification and training requirements because they offer a different model than the United States. Specifically, individuals with no previous flying experience are generally trained from the beginning to be commercial airline pilots. The site visits allowed us to obtain information on European countries' pilot standards, as well as their efforts to revise their piloting performance and training standards to competency-based training models. During these site visits, we interviewed officials at the European Aviation Safety Agency (EASA) (Europe's aviation regulatory organization), civil aviation authority officials, and representatives from international and European aviation stakeholder groups, airlines, and pilot training schools.

We conducted this performance audit from March 2010 through November 2011 in accordance with generally accepted government auditing standards. Those standards require that we plan and perform the audit to obtain sufficient, appropriate evidence to provide a reasonable basis for our findings and conclusions based on our audit objectives. We believe that the evidence obtained provides a reasonable basis for our findings and conclusions based on our audit objectives. See appendix I for more information about our scope and methodology.

BACKGROUND

Civil aviation in the United States can be generally divided into two broad categories—general aviation and commercial aviation. General aviation comprises all aviation activities other than military and commercial airlines. All civilian students are trained in the general aviation sector until they are hired as airline pilots. Commercial aviation generally refers to businesses that carry passengers or cargo for hire or compensation. To operate as a commercial airline, a business must have an airline operating certificate issued by FAA, based on federal aviation regulations, which is determined by the type of commercial service being provided.[6] Airlines that provide scheduled commercial service are often grouped into two categories. Mainline airlines are the traditional large airlines that provide domestic and international passenger service on larger aircraft such as American Airlines or Delta Airlines. Regional airlines, such as Mesa and Piedmont Airlines, also provide domestic and international passenger service, generally using aircraft with fewer than 90 seats and serving smaller airports. The international service that regional airlines provide is confined to border markets in Canada, Mexico, and the Caribbean. More than 13,000 regional airline flights operate daily, which represents more than half of the number of U.S. domestic flights.

As the federal agency responsible for regulating the safety of civil aviation in the United States, FAA is responsible for the administration of pilot certification (licensing) and conducting safety oversight of pilot training. Regulations for pilot certification and training are found in three different parts of the Federal Aviation Regulations—Parts 61, 141, and 142.[7] All pilots are subject to a series of certification requirements established by FAA, but the requirements vary depending on the type of training environment.[8] Part 61 recognizes six basic types of pilot certification: student, sport, recreational, private, commercial, and airline transport.[9] Part 61 also establishes the core training requirements for each pilot certification, which describes the eligibility requirements, aeronautical knowledge and flight proficiency standards, and the required flight hours (see table 1). Pilot training can be provided to students by flight instructors under Part 61.[10] Part 141 outlines the specified personnel, aircraft, facilities, curriculum, and other operating requirements that approved pilot training organizations (schools) must meet in order to hold an operating certificate from FAA. Part 142 outlines specific requirements for training centers that primarily relate to the advanced training provided to pilots by employers, such as airlines. Our report focuses on the initial pilot training that students are provided until they are hired as airline pilots; thus, advanced training for pilots is not within the scope of our study.[11]

To obtain a private, commercial, or airline transport pilot certificate from FAA to perform various aviation activities, individuals typically have to successfully complete pilot training and pass the following two FAA tests for each pilot certificate and rating obtained:[12]

- A knowledge test assesses applicants' understanding of the aeronautical knowledge areas required for a specific certificate or rating and can be administered in written form or by a computer.
- practical test consists of a flight test and an oral examination. The flight test assesses applicants' knowledge of the areas of operations of an aircraft and the ability to demonstrate the maneuvers in an aircraft while in flight. The oral examination is conducted by having an applicant respond to random questions related to aviation knowledge

and aircraft operations before, during, and after the flight test, and typically lasts between 1 and 2 hours.[13]

Table 1. Training and Total Flight-Hour Requirements for FAA's Three Main Pilot Certificates—Private, Commercial, and Airline Transport Pilot

Type of pilot certificate	Part 61 required ground school training hours for aeronautical knowledge	Part 61 required flight hours	Part 141 required ground school training hours for aeronautical knowledge	Part 141 required flight hours
Private pilot	No minimum hours specified; minimum knowledge subject areas must be covered	40	35	35
Commercial pilot	No minimum hours specified; minimum knowledge subject areas must be covered	250	35	190
Airline transport pilot	No minimum hours specified	1,500	40	1,500

Source: GAO presentation of information contained in Part 61 and Part 141of Title 14 of the U.S. Code of Federal Regulations.

To become a certified commercial pilot,[14] which is currently the minimum requirement for being hired by an airline as a first officer, individuals also must undergo several steps of pilot training and certification in accordance with FAA regulations.[15] Once cleared by the medical examination, students obtain a medical certificate and a student pilot certificate from FAA.[16] Figure 1 shows the typical progression of training and certifications required to become an airline pilot.

Figure 2 shows examples of the progression from single-engine trainer, to multi-engine (i.e., jet) trainer—used by some pilot schools to provide students with the multi-engine rating—to the much larger, faster jet used by regional airlines. Once commercial pilots complete the process of initial training, they are qualified to apply for a first officer pilot position at an airline. Entry-level positions are typically at regional, and not mainline, airlines; mainline airlines typically draw from regional airlines for their pilots. If hired, pilots must complete the airline's new hire training, which consists of indoctrination, ground and aircraft systems, simulator training, and the initial operating experience, wherein the pilot applies what they learn in the previous training phases. The airline submits these training programs for approval by the FAA to ensure they meet Part 121 requirements.

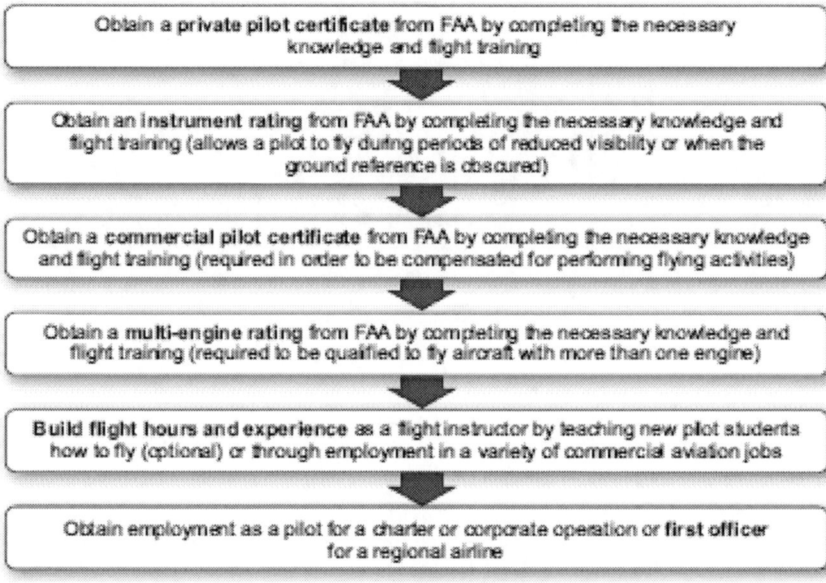

Source: GAO.

Figure 1. The Typical Path to Become a Commercial Airline Pilot.

As part of its oversight responsibility, FAA has safeguards in place to ensure that pilot applicants are provided the necessary training and undergo complete and thorough pilot certification examinations. The National Program Guidelines (NPG), initiated in 1985, are oversight policy guidelines developed annually by FAA for its eight regional offices and their associated local district offices for oversight of pilot schools, pilot examiners, and flight instructors.[17] The NPG identifies required inspections and optional inspections.[18] As part of this oversight process, FAA uses the Program Tracking and Reporting Subsystem (PTRS) for scheduling and recording inspection records and findings for NPG inspections of pilot schools, flight instructors, and pilot examiners.[19] Additionally, FAA uses the Enforcement Information System (EIS) for tracking and reporting information about any enforcement actions the agency takes for statutory or regulatory violations.[20]

As a member of the International Civil Aviation Organization (ICAO), the United States conforms to international standards and recommended practices for pilot training and certification.[21]

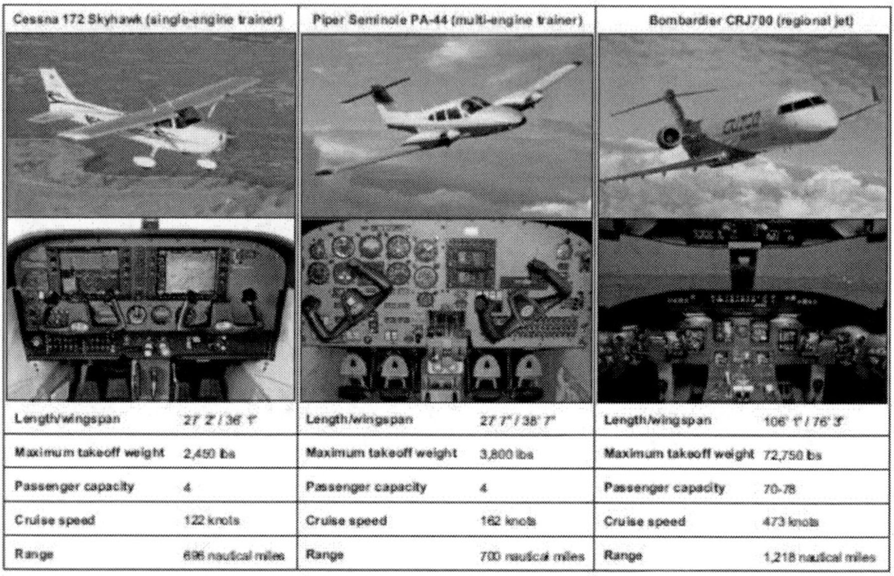

Sources: Cessna, Piper and Bombardier.

Figure 2. Examples of Single- and Multi-Engine Aircraft Used by U.S. Pilot Schools and a Multi-Engine Jet Used by Regional Airlines.

ICAO is the international body that, among other things, promulgates international standards and recommended practices in an effort to harmonize global aviation standards. These standards and recommended practices are developed to ensure that civil aviation throughout the world is safe and secure. ICAO has no enforcement powers and only establishes recommended standards and guidelines, e.g., licensing requirements for flight crew personnel, including pilots. Therefore, ICAO members (known as contracting states) decide whether to incorporate the standards and recommended practices into national laws or aviation regulations.

Pilot Schools' Training Varies, and Although All Students Must Pass the Same Test, Concerns Exist Related to the Quality of the Testing and Other Training Requirements

The roughly 3,400 U.S. pilot schools can be divided into three categories: (1) non-collegiate flight instructor-based schools, (2) non-collegiate vocational pilot schools, and (3) collegiate aviation schools. The training provided by the school varies in the minimum requirements for the flight and ground school hours required for each certification level, level of oversight provided by FAA, and level of educational instruction being provided by type of school. However, all student pilots have to successfully complete ground and flight training and pass the same knowledge and practical tests prior to receiving a pilot certificate from FAA. Nevertheless, there is no consensus and little empirical evidence on how the different pilot training schools compare in preparing professional pilots for the commercial airline industry. In addition, modern aircraft used by regional airlines have evolved and the operational demands have increased on pilots in high-altitude and complex airline operations; yet, U.S. pilot training requirements for certification of commercial pilots were last revised in 1997. For example, it is possible for an individual to obtain all levels of pilot certifications (i.e., private, commercial, and airline transport) in a general aviation flying environment in a single-engine aircraft. However, in order to qualify to be hired by an airline, a commercial pilot would also need to obtain instrument and multi-engine ratings. Some stakeholders, including representatives of regional airlines, we interviewed said the current training regulations for commercial pilots should be revised to incorporate additional training that would improve the competency of entry-level first officer applicants. FAA has initiated several efforts to address issues related to pilot training and certification testing. Recent legislation requires that FAA develop regulations increasing pilot certification requirements for all airline pilots.

The Three Different Types of Pilot Training Schools Vary in Type of Training, but All Prepare Students to Take the Same FAA Tests

Approximately 3,400 pilot schools exist in the United States and the most basic difference among the types of schools is the training environment provided to the students.

For the reporting purposes of our study, we divided them into three categories: (1) non-collegiate flight instructor-based schools, (2) non-collegiate vocational pilot schools, and (3) collegiate aviation schools.

- Non-collegiate flight instructor-based schools (Part 61). Pilot training conducted under Part 61 regulations is often provided by an individual, for-hire flight instructor who can operate independently as a single-instructor school at a local airport with a single aircraft on which to train students. Other flight instructor-based schools operate as a more traditional training school with a small, physical facility located at a particular airport. These schools are the most common type (see Figure 3). The majority of students that complete training in non-collegiate, flight instructor-based schools are generally interested in recreational flying, although most commercial pilots in the United States also undertake this type of training as the initial path toward becoming an airline pilot. Flight instructor-based schools offer flexible training environments to meet specific students' needs as long as they pass the final tests. These schools are not subject to direct FAA oversight beyond the initial certification and subsequent renewal of the flight instructor's certificate. However, flight instructors may be inspected by FAA when a triggering event occurs regarding the training being provided, such as being involved in an aircraft accident.

- Non-collegiate vocational pilot schools (Part 141). Vocational schools elect to apply for an operating certificate from FAA to provide pilot training under Part 141 regulations. Part 141 regulations require these schools to meet prescribed standards with respect to training equipment, facilities, student records, personnel, and curriculums.22 Vocational schools must have structured and formalized programs and have their detailed training course outlines or curriculums approved by FAA. Curriculums can vary in content, but FAA provides fundamental core training guidelines that must be followed within the curriculum for the school to receive a certificate.23 These schools do not allow the flexibility of flight instructor-based schools as the training sequence outlined in the curriculum cannot be altered. FAA requires annual inspections of these schools, unlike flight instructor-based schools.
- Collegiate aviation schools (Part 61 or Part 141). Pilot training is also provided through 2- and 4-year colleges and universities, which typically offer an undergraduate aviation-based degree along with the pilot certificates and ratings necessary to become a commercial pilot. In general, most of the collegiate aviation schools provide pilot training under a Part 141 certificate, although they can provide training under Part 61. Collegiate schools that provide training under Part 61 regulations generally offer similar structured, curriculum-based training as collegiate schools with a Part 141 certificate.

Figure 3 displays how each of the three types of pilots schools are dispersed across the United States.

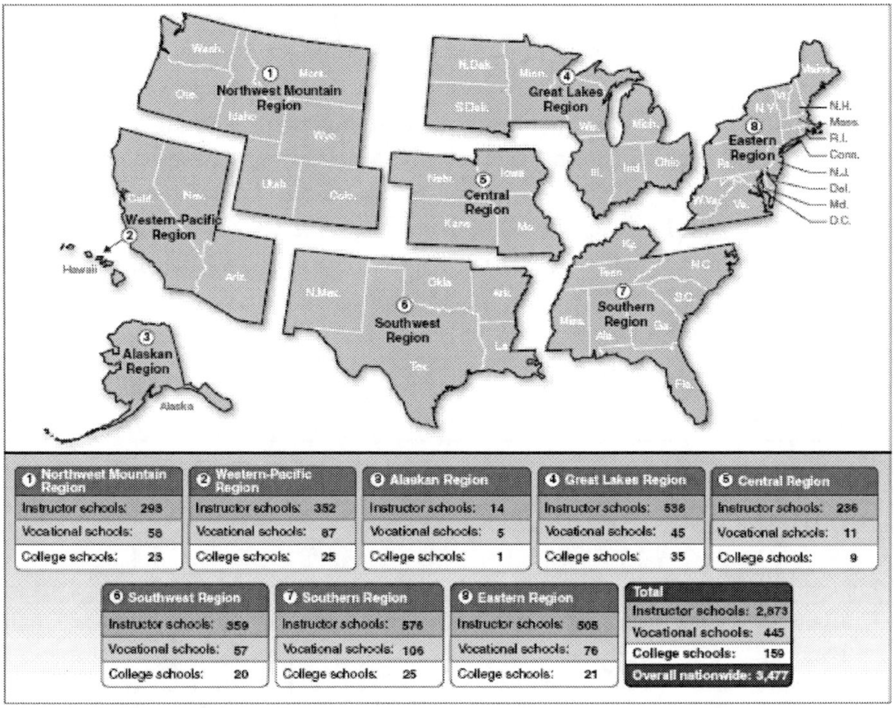

Source: GAO analysis of FAA and Aircraft Owners and Pilots Association information; Map Resources (map).

Note: The total number of schools is approximate because a comprehensive list for the various types of schools does not exist. The total number of schools was provided by the Aircraft Owners and Pilots Association (AOPA) as of June 2010. Association officials noted that information is entered into the listing database by the schools and could not guarantee accuracy and the currency of the data. FAA maintains only a current list of the pilot schools with a Part 141 operating certificate. Since most of the pilot schools are not directly tracked by FAA, the AOPA listing is considered the best available. We reconciled the AOPA total list of schools with FAA's current list of pilot schools with a Part 141 operating certificate as of June 2011. We eliminated the duplicate schools and added schools from the FAA list to the total number of schools.

Figure 3. The Number and Types of Pilot Schools across the Eight FAA Regions.

For the most part, all pilot schools must provide training that includes both classroom and flight training.[24] Classroom training, or ground school, provides students with the required aeronautical knowledge and cognitive skills necessary to perform the tasks required to become a pilot. Flight training focuses on learning how to manipulate the controls of an airplane and make it perform certain maneuvers. Regardless of the type of school, flight instructors must teach students by demonstrating and explaining, on the ground and in the air, the basic principles of flight (e.g., airspace, aerodynamics, weather factors, and Federal Aviation Regulations). The number of training flight hours required for pilot certification varies by the aviation regulations being used to provide pilot training. Because training under Part 141 regulations requires a school to use an FAA-approved curriculum, fewer hours of actual flight training are required than under Part 61.[25] Figure 4 shows the differences in general characteristics of the types of U.S. pilot schools.

Type of pilot school	Ground school and flight training	Knowledge test	Practical test	FAA-approved curriculum	Degree granted	Accreditation
Flight instructor-based (part 61)	✓	✓	✓			
Vocational (part 141)	✓	✓	✓	✓		
Collegiate (part 61 or 141)	✓	✓	✓	✓[a]	✓	✓

Source: GAO.

Note: Although a training curriculum does not require FAA approval under Part 61, flight instructors must cover the minimum knowledge subject areas outlined in the regulations.

[a] Not all collegiate aviation schools have a Part 141 certificate. Therefore, for collegiate schools that provide training under Part 61, no FAA approval is needed for the training curriculum used for training.

Figure 4. Differences in Regulatory Requirements and General Characteristics of U.S. Pilot Schools.

FAA regulations do not prescribe the entry requirements, selection criteria, and screening procedures for students seeking entry into U.S. pilot schools, and as a result, they could vary considerably among schools. In general, pilot schools admit those students who can pay for the training; however, FAA sets a minimum age requirement for each pilot certification and requires a current FAA medical certificate and that every pilot student is able to read, speak, write, and understand the English language.[26] FAA's long-standing requirement for English proficiency complies with the 2008 ICAO standard that all private, commercial, or airline transport pilots who operate internationally have a pilot certificate with the level of English language proficiency. If a person is determined to be proficient during the FAA practical test for pilot certification, FAA issues pilot certificates with an "English Proficient" endorsement to attest that the pilot meets the ICAO standard.

One of the distinctive characteristics of collegiate schools is that they are generally accredited academic programs, which recognizes a level of program quality.[27] However, a recently-created organization, the Flight School Association of North America, implemented an accreditation program in August 2011 for non-collegiate pilot schools intended to establish an educational quality standard.[28] According to association officials, accrediting non-collegiate pilot schools will help to level the playing field with the collegiate schools and assist consumers in comparing pilot schools.

Regardless of the type of pilot schools that students attend, once training has been completed pilot candidates must pass the same knowledge and practical tests to obtain a pilot certificate.[29] FAA uses a multiple-choice, knowledge test to measure the extent to which applicants for FAA pilot certificates have mastered the required aeronautical knowledge areas provided in ground school. To pass, applicants must achieve an overall score of 70 percent or higher. However, concerns have been voiced by some aviation stakeholders related to whether the current knowledge test actually requires students to learn the material, as opposed to simply studying sample test questions from publicly available sources. Literature related to pilot certification and training issues and some aviation stakeholders have pointed out that FAA testing is generally based on rote memorization. They stated that this encourages instructors and students to focus on memorizing test questions to pass the required FAA knowledge test, rather than developing a true understanding of the material. In 2004, the National Aeronautics and Space Administration (NASA) published a study on FAA's pilot knowledge tests.[30] NASA found that many applicants completed the test in far less time than would be required for the average human to even read the questions and answers on the test—indicating that students had memorized the questions and answers—which raises concerns about the extent to which students actually mastered the material.

Little Information Exists for Examining and Comparing How the Different Types of Pilot Schools Prepare Pilots for Airline Operations

There is little empirical research that has been conducted comparing the extent to which different types of pilot schools prepare pilots for the commercial airline industry. We reviewed a pilot source study, published in 2010, authored by professors from several accredited collegiate aviation schools that researched the impact of collegiate aviation training on preparing students to be regional airline first officers.[31] The researchers analyzed data on how 2,156 new-hire pilots performed in the training programs of six regional airlines from 2005 through 2009. The study found that the new-hire first officer pilots with the highest rate of success—in terms of the amount of extra training needed to complete training tasks and fewer number of times tasks were not completed—in the airline training: (1) were graduates of accredited college flight degree programs, (2) had experience as flight instructors, and (3) had accrued between 500 and 1,000 flight hours. We reviewed the study and determined that, while statistically significant, the results of this research showed small differences in the correlations that supported these three conclusions. Phase II of the study, completed in early 2011, expanded the research to include testing of multiple variables with the same dataset, but the researchers did not report the results of the tests they conducted of their data. Phase III is currently underway and is expanding the current dataset to include more than six regional airlines. Phase IV will include more detailed background data on the newly hired pilots to determine relationship factors.

Other than the 2010 pilot source study, our literature review found that little other academic study exists and there is no consensus about how well the different types of pilot schools prepare commercial pilots for airline operations. We received a variety of perspectives from industry representatives and some anecdotal information that suggested that one major benefit to completing a structured training program that is the training provides better aeronautical knowledge (ground school) than an unstructured learning environment. Officials from two regional airlines, two collegiate aviation schools, and four industry associations with whom we spoke generally agreed with the results of the initial phase of the pilot source study. Officials from the Regional Airline Association (RAA) told us that the broad-based curriculum used by vocational and collegiate aviation schools is the better method for preparing a person for a professional airline career.[32] Furthermore, officials from 6 of the 12 industry associations and one mainline airline we interviewed considered the quality of education at many of the collegiate aviation schools to be the highest level of civil aviation pilot training because collegiate schools are designed to produce professional pilots for airlines, rather than for other aviation jobs.

Collegiate curriculums also cover a broad range of areas above FAA minimum training requirements. In addition, representatives from all but one of the regional airlines we interviewed generally told us they strongly preferred, but do not require, first officer candidates trained in collegiate aviation schools because they perform better in their airline's training program when hired.[33] Due to limited screening, training structure, and variability of educational content, according to some of the regional airline officials, flight instructor-based schools are less likely to produce first officers that are prepared immediately upon completing the training to enter the workforce and succeed in an airline environment. On the other hand, stakeholders from 4 of the 12 industry associations pointed out the large number of pilots that matriculate through flight instructor-based schools and many are hired by regional airlines without any performance issues. Representatives from two of the regional airlines indicated that the professional pilot experience gained through commercial aviation positions after completing pilot training is more important than the type of pilot school attended.

Pilot Training Requirements for a Commercial Pilot Certificate Are Not Aligned with Airline Operations

Several industry stakeholders have stated that current training requirements for commercial pilots are not aligned with today's commercial airline environment. FAA requires the same initial training for a pilot hired as a first officer of a regional airline carrying passengers as it does for a pilot hired to fly for a commercial non-airline operation, for example crop dusting.[34] The Air Line Pilots Association (ALPA) has suggested that FAA revise the regulations to make a clear distinction between training and certification requirements for airline operations and those for other types of commercial operations. ALPA contends the regulations were developed in an era in which commercial pilots were hired by airlines in small, slow, propeller-driven aircraft or as flight engineers on jet-powered aircraft. It would traditionally take several years and thousands of flight hours before these pilots were given an opportunity as a first officer of jet transports.[35] However, according to ALPA, it is not uncommon today for newly hired pilots to be hired directly into airline training programs to become first officers of high-altitude, complex aircraft carrying 50 or more passengers, the type of aircraft that warrants pilots with more knowledge and greater skills than the new-hire airline pilots of the past. Officials from two industry associations and eight regional airlines advocated for a separate pilot certification track with additional training requirements specific to being an airline pilot. Because airline pilots are responsible for the safety of the flying public, according to ALPA, it is reasonable that they should be held to a higher standard of competency, knowledge, and training than pilots in other flight operations.

Additionally, requirements for a commercial pilot certificate do not emphasize training in some areas—like decision-making and using modern technologies—that are directly related to the airline pilot profession.

- According to FAA and other stakeholders, the regulations regarding ground school and flight training, as well as the test standards for a commercial pilot certificate, generally emphasize the mastery of maneuvers and individual tasks to determine competence. The emphasis is on development of motor skills to satisfactorily accomplish individual maneuvers—whereas only limited emphasis is placed on decision-making—unlike in scenario-based training that emphasizes improving operational experience.[36]
- In addition to traditional skills of flying, navigating, and communicating, pilots in today's newer aircraft have to manage automation, information displays, and other new technologies. According to the FAA Industry Training Standards' guidance material for the commercial pilot certificate, a growing number of pilots are being hired by regional airlines as first officers to operate aircraft with these advanced avionics and systems.[37] While these pilots may gain flying experience and spend years building flight time in commercial non-airline jobs or as flight instructors, this experience may be accumulated in smaller, slower, and less advanced aircraft. Modern aircraft offer advanced avionics and performance capabilities and many of these new aircraft travel faster and further than older generation commercial aircraft. While generally considered enhancements, these modern technologies require increased technical knowledge of newer systems and avionics and new skills for managing automation and computerized flight and navigation systems.

According to literature, as airspace complexity and air traffic density increase, airline pilots must have increased situational awareness, understand risk assessment, and have a complete understanding of managing the automation of the aircraft.[38] The current training requirements and testing for a commercial pilot certificate do not emphasize the development of these skills.

Representatives from 10 regional airlines, 4 pilot schools, and 2 industry associations we interviewed said the current training regulations for commercial pilots should be revised to incorporate additional training requirements that would improve the performance capabilities of the first officer applicants that seek employment at airlines, such as exposure to advanced jet concepts and simulation,[39] aircraft unusual attitude (i.e., upset and stall recovery),[40] flight crew coordination and environment,[41] and scenario-based training. However, when pilots are hired by airlines, these types of training are provided by the airline to ensure that pilots are adequately competent in these and other advanced training areas—some required by FAA for airline operations. For example, FAA regulations for airline operations require that all pilots are provided crew resource management training as part of the airline's new hire and recurrent training programs. According to ALPA, the lack of specific training requirements to be a commercial airline pilot results in a wide range of initial training experiences, not all of which are well suited for the commercial airline industry. To compensate, some regional airlines, such as SkyWest Airlines, use various flight training devices to screen pilots during the hiring process to gauge their piloting skills (see Figure 5). However, if additional training is required by FAA for pilots prior to being hired by an airline, the students would likely be responsible for the extra costs involved and would add to the total costs of pilot training borne by the student. (For more information on the costs associated with pilot schools, see app. II.)

The industry concerns about current training regulations for commercial pilots and incorporating additional initial training requirements to improve first officer applicants' performance capabilities could be addressed by the Airline Safety and Federal Aviation Administration Extension Act of 2010 for all airline pilots.[42] Currently, while a captain for a commercial airline is required to hold an airline transport pilot certificate—the highest level of pilot certification and requiring the highest number of total flight hours—a first officer is required to hold only a commercial pilot certificate, i.e., requiring a minimum of 250 flight hours.[43] However, the recent law will require that each pilot (captain and first officer) must have an airline transport pilot certificate, which currently requires a minimum of 1,500 total flight hours.[44] Individuals interested in becoming a first officer for a regional airline generally complete training from pilot schools with a commercial pilot certificate and possess about 300 to 500 flight hours. The 2010 law directs FAA to conduct a rulemaking and effect the changes no later than August 2013.[45] According to FAA, it will issue a notice of proposed rulemaking regarding the increased requirements in the fall of 2011.

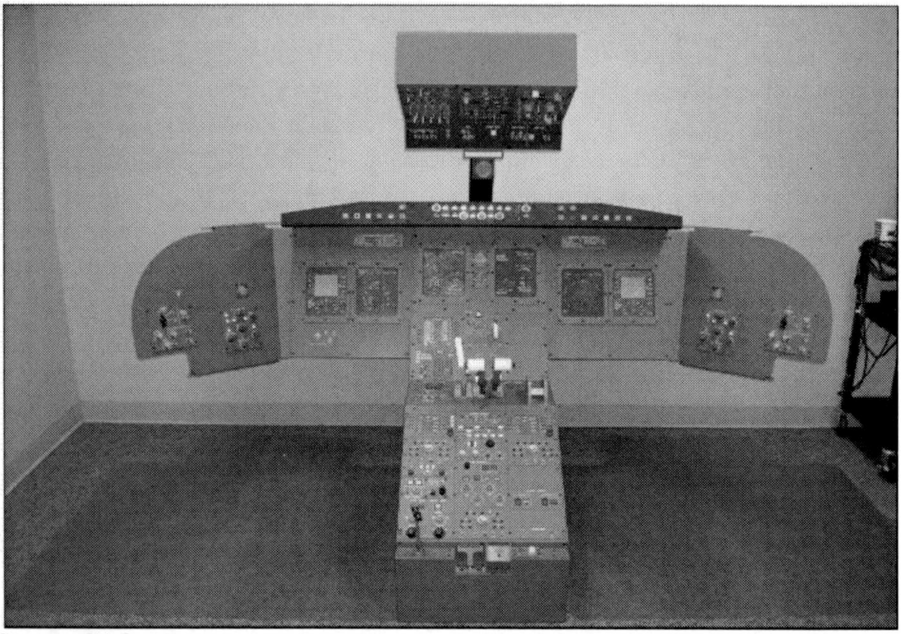

Source: GAO.

Figure 5. SkyWest Airlines' Flight Training Device Used to Test Pilot Applicants.

Representatives of the regional airlines we interviewed were concerned this legislation will reduce airlines' hiring flexibility and make it harder to find qualified first officers that possess an airline transport pilot certificate.

As another potentially relevant factor, the law stated that the FAA Administrator may allow specific academic training courses—beyond the additional courses required by the legislation to raise the minimum requirements for the airline transport pilot certificate—to be credited in lieu of flight hours needed to obtain an airline transport pilot certificate. According to FAA's First Officer Qualifications Aviation Rulemaking Committee Report,[46] well-structured training programs that feature integrated academic content and flight experience optimize the pilot learning process, and the committee supported new, higher-level minimum certification requirement for first officers. To support the concept that academic training courses should be credited for some of the additional total flight hours, the report outlined a system for crediting academic training courses based on the sources of pilot training, e.g., vocational pilot schools, collegiate aviation schools, or military. The Coalition of Airline Pilots Associations (CAPA) and National Air Disaster Alliance-Foundation presented dissenting opinions to this approach in the report and suggested that academic courses, while necessary, should not replace an increase in total flight hours required in the law.[47]

Some Pilot Schools Provide Training That Is More Relevant to Airline Operations

Many of the collegiate aviation schools provide specialized training in a flight simulation training device using realistic scenarios, including some coursework and advanced flight training in jet aircraft systems and airline operational procedures.

For example, Embry-Riddle Aeronautical University's Aeronautical Science degree program is designed to prepare graduates for a career as a professional pilot in multi-crewmember, jet aircraft.

Courses include communication theory and skills, aircraft turbine engines, crew resource management, aviation weather, jet transport systems, and optional upset recovery training. Officials who represented 10 of the 24 regional airlines we interviewed listed some of these types of courses as examples that FAA could require as part of pilot schools' training curriculums that would improve the skill level and competency of applicants seeking to be hired as first officers.

Some collegiate aviation schools—and some large flight instructor-based and vocational pilot schools—have developed relationships with the training departments of some regional airlines, referred to as bridge programs, in order to qualify their students with advanced training procedures involving regional jet simulators.[48] Students enrolled in a bridge program will train on flight simulators and other flight training devices, become familiar with regional jets, and often learn airline-specific operational procedures in a multi-crew environment.

The airlines, in turn, will offer interviews to students from those programs that successfully complete the school's curriculum and earn all of the pilot certification credentials.

A bridge program is designed to bridge the gap between the general aviation training experience in small single- and multiengine aircraft and a professional airline career.

For example, Arizona State University's aviation degree program has established a bridge agreement with a regional airline, Mesa Airlines, which allows the students to train in full-motion simulators of the regional jets that Mesa operates (see Figure 7).

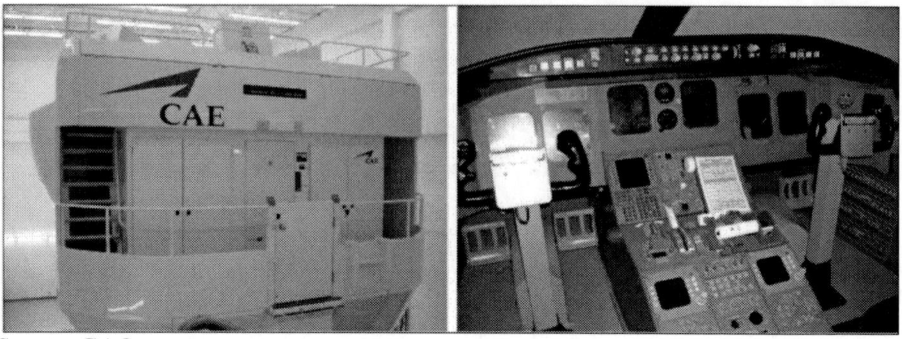

Source: GAO.

Figure 6. Full-Motion Flight Simulator at Arizona State University.

FAA and Industry Efforts to Address Issues Associated with Pilot Training

FAA has initiated several efforts to address issues related to pilot certification testing and training. In 2010, FAA began revising the repository set of questions that it uses to create its knowledge tests for pilot certification. FAA found that a significant percentage of applicants tested on the new questions failed the test compared to those that took the test with the previous questions. FAA plans to cooperate with industry representatives on future changes to the knowledge test questions and would likely implement any further changes over the next 2 years. Additionally, according to FAA officials, FAA has developed other plans related to improving pilot training to be implemented during fiscal year 2012.

- FAA plans to establish a government and industry working group during fiscal year 2012 to address issues related to pilot certification testing standards and training. The group will make recommendations to FAA on a variety of issues, including knowledge content, technical information related to pilot knowledge and practical tests, computer testing supplements, knowledge test guides, and practical test standards.
- FAA is currently updating its national guidance and associated handbook for FAA inspectors on the recurrent training for flight instructors that is conducted by the aviation industry by providing refresher training courses. As stated earlier, part of the oversight for

flight instructor-based schools is the subsequent renewal of the flight instructor's certificate every 24 months. The refresher training is designed to keep flight instructors informed of changes to flight training and is one of several methods by which a flight instructor may renew a flight instructor certificate.[49]

- FAA is also updating its guidance on the review process for the 24-month certification renewal for pilots and flight instructors. According to FAA, the current guidance is outdated and the revised version will provide more detailed guidance for the renewal review process, updated terminology and references, and reorganized review content.
- FAA plans to make changes to the practical test standards (i.e., guidance used for conducting the flight test portion of a practical test) to incorporate required testing for runway incursions. FAA's goal with the revised standards is to reduce runway incursions by 2 percent annually from the current level.50 Currently, the part of the practical test standards for evaluating a student's knowledge of runway incursions is not required or specifically outlined. However, changes to the practical test standards include labeling runway incursions as a required testing task that includes specific procedures to be conducted during the test.
- FAA has also recently initiated efforts to partner with the aviation academic community through a 5-year plan initiative, working through AABI and the University Aviation Association (UAA), to leverage academic expertise and develop best practices for improving all pilot training.51 The goal is to identify specific non-regulatory measures that can be used to improve training and reduce accidents.

Other efforts to improve pilot training have generally focused on advanced training for pilots already employed at airlines—not on initial pilot training for students at pilot schools. Over time, U.S. pilot schools have become the primary source for producing pilots for the airline industry. In May 2011, in response to comments to a January 2009 notice of proposed rulemaking and requirements in the Airline Safety and Federal Aviation Administration Extension Act of 2010, FAA issued a supplemental notice of proposed rulemaking that would require existing airline pilots to train as a complete flight crew and coordinate their actions through crew resource management and scenario-based training, among other things.[52] The rule would require airlines to provide ground and flight training to all existing airline pilots in the recognition and avoidance of stalls, recovery from stalls, recognition and avoidance of aircraft upset, and the proper techniques to recover from upset—all related factors evidenced in the Colgan Air crash in 2009—as required in the act.[53] Additionally in May 2011, NTSB issued a series of recommendations to FAA related to first officer leadership and communications training. One recommendation to FAA, to which FAA has not yet formally responded, was that role-playing or simulator-based exercises that teach first officers to assertively voice concerns and that teach captains to develop a leadership style that supports first officer assertiveness be included as part of the already required crew resource management training for airline operations. These proposals are designed to enhance the training programs of airlines rather than the training requirements for pilot students, however, some of these ideas could be applied to the initial training for students seeking commercial airline careers.

In addition, other aviation industry efforts are being developed that focus on improving all pilot training. For instance, one aviation organization has proposed the development of a global professional pilot certification to bridge the competency gap between pilot certification and being employed. The professional pilot certification would be based on a set of standards to ensure a newly-certified pilot has the knowledge required by the standards to enter the pilot profession. A not-for-profit association of pilot training professionals has proposed the development of an independent, international clearinghouse for pilot training best practices. International Air Transport Association (IATA) launched the Training and Qualification Initiative in 2007 to update and modernize the training of existing and future pilots. The initiative's goal is to make pilot training more accurately reflective of the needs of flight deck operational procedures. Aviation stakeholders, such as Boeing and Flight Safety International, are offering a first officer's course to bridge the skill and competency gap between training received at pilot schools and being employed as a commercial airline pilot.[54]

European Pilot Training Differs from the U.S. System, in Part Due to Varying Philosophies and Circumstances

Similar to FAA, EASA provides the regulatory framework for oversight of European countries' national aviation authorities, which carry out the requirements for pilot licensing and training.[55] Creation of EASA is the centerpiece of the European Union's strategy for developing one level of aviation safety across Europe. EASA establishes common safety and environmental regulations and standards, and monitors the implementation of standards through inspections in its member states under authorization of the European Union. The United States and Europe provide a good comparison for aviation licensing as they offer equivalent pilot certificates, but they provide training for these certificates in different ways, as we describe below.

The European Pilot Training System Emphasizes Aeronautical Knowledge (Ground School) Instruction and Testing While the U.S. System Focuses More on Training in Practical Skills

Both FAA and EASA meet the ICAO standards for pilot certification and offer the private pilot, commercial pilot, and the airline transport pilot certifications (see table 2). An FAA pilot certificate is the equivalent of an EASA pilot license (we will only refer to a certificate—not license—for the simplicity of reporting).

Table 2. Comparison of Pilot Certifications and Training Standards of ICAO, FAA, and EASA

Private Pilot Certificate			
Minimum Age	ICAO	FAA	EASA
	17	17	17
Ground school	No specified minimum number of hours	- No specified minimum number of hours for Part 61 - 35 hours for Part 141	No specified minimum number of hours
Written test	No reference to a test, just demonstrate knowledge	Pass an examination and score at least 70 percent overall	Pass examination and score at least 75 percent for each subject tested
Flight training	40 hours total time; 35 hours total time if completed during a course of approved training	- 40 hours (Part 61) or 35 hours (Part 141) total time	45 hours total time; 5 hours may be completed in a Basic Instrument Training Device or a flight simulator
Practical test	Demonstrate ability as pilot-in-command of an aircraft within the appropriate category of aircraft	Required in aircraft	Required in aircraft

Table 2. (Continued)

Private Pilot Certificate			
Minimum	ICAO	FAA	EASA
Age	17	17	17
Commercial Pilot Certificate			
Minimum age	18	18	18
Ground school	No specified minimum number of hours	- No specified minimum number of hours for Part 61 - 35 hours for Part 141	- 250 hours (modular)a - 350 hours (integrated)b
Written test	No reference to a test, just demonstrate knowledge	Pass an examination and score at least 70 percent overall	Pass examination and score at least 75 percent for each subject tested
Flight training	200 hours or 150 hours total time if completed during an approved training course	250 (Part 61) or 190 hours (Part 141) total time	200 (modular) or 150 hours (integrated) total time
Practical test	Demonstrate ability as pilot-in-command of an aircraft within the appropriate category of aircraft	Required in aircraftc	Required in aircraft
Airline Transport Pilot Certificate			
Minimum age	21	23	21
Ground school	No minimum number of ground school hours	- No specified minimum number of hours for Part 61 - 40 hours for Part 141	- 650 hours (modular) - 750 hours (integrated)
Written test	No reference to a test, just demonstrate knowledge	Pass an examination and score at least 70 percent overall	Pass examination and score at least 75 percent for each subject tested
Flight training	1,500 hours total time	1,500 hours total time	1,500 hours total time
Practical test	Demonstrate ability as pilot-in-command of an aircraft within the appropriate category required to be operated with a copilot	Required in aircraft or flight simulator	Required in aircraft or full flight simulator of a multi-pilot aircraft under instrument flight rules as a pilot-in-command

Source: GAO presentation of ICAO standards, and FAA and EASA regulations.

[a] Modular training is similar to the training model practiced in the United States under Part 61 where students are trained through a building block approach of modular segments of ground school and flight training at their own pace.

philosophy of *ab initio* training—which is also the traditional approach by which U.S. military pilots are trained—is the belief that a competent, proficient airline pilot can be trained to airline standards with as little as 350 hours of flight time, provided the student is immersed in a properly designed aviation curriculum from the outset.[58] In other words, the training approach is not based on the quantity of hours of training, but rather on the quality of the training to better enable an individual to achieve competency. A few U.S. pilot schools offer an *ab initio* training program; however, they are generally provided by universities and require the student to take part in a 4-year program, generally a longer timeframe than it takes to complete a European *ab initio* program administered by an airline sponsored school. U.S. *ab initio* programs train pilots for positions with regional or commuter airlines, whereas in Europe, *ab initio* training is more specifically for pilots being training for mainline airlines.

Another major difference between the United States and European countries is the implementation by EASA of regulations to support the multi-crew pilot license—an ICAO-approved, alternative pilot-training and certification concept specifically geared toward training commercial airline pilots.[59] According to ICAO, a total of 32 ICAO member states have regulations in place for the multi-crew pilot license. However, currently only 13 of the 190 ICAO members (7 percent) have approved training organizations to conduct training for the multi-crew pilot license, with different training schemes in progress. The training is designed to focus on mastering the competencies specific for becoming an airline first officer. The multi-crew pilot license, established by ICAO, requires at least 240 hours of total flight training and is comprised of actual flight time and simulation time for meeting competency milestones. However, the license is not a general pilot certificate and must be granted to an individual for a specific aircraft type and limits an individual as a first officer for a specific airline. FAA has not developed regulations for a multi-crew pilot certification, and there are differing views on its usefulness and necessity in the United States. FAA officials said they have been studying the feasibility of implementing the necessary regulations for U.S.-based commercial airlines, but they also indicated that U.S. airlines have not publicly shown interest in a multi-crew pilot certification due to the availability of a broad pool of commercial and airline transport pilots in the United States. Representatives from three regional airlines and one industry association told us that, with the number of furloughed pilots as a result of the economic downturn in 2008, application of the multi-crew pilot certification is not needed in the United States and would be too restrictive in nature. The certification would limit pilots to being first officers, limit them to a specific aircraft type, and not allow them to transfer to other airlines. Traditional pilot certificates require more training hours, but do not include such restrictions.

Part of the reason European countries and airlines have adopted the *ab initio* approach is to address a shortage of qualified airline pilots. Historically, U.S. airlines have recruited experienced pilots from the robust U.S. general aviation community and the U.S. military. The United States also has significantly more pilot schools than Europe. Conversely, Europe has not benefited from a steady stream of military pilots or a thriving general aviation sector. As a result, European schools are mostly focused on producing commercial airline pilots. European countries and airlines have used the *ab initio* training model and multi-crew pilot license to increase the number of available airline pilots.

At times, in response to pilot shortages, European airlines have funded the training for their pilot candidates. After the screening process, many student pilots in the European countries we visited are provided training by airline sponsorship or an agreement for employment with an airline. Examples of airlines that follow this practice include Lufthansa and Air France, where students are offered the training as part of a partial sponsorship program, wherein candidates are required to pay a small portion of the training costs upfront (Lufthansa provides loans to students to cover this cost). Once training is completed, Lufthansa and Air France enter into an employment contract with the candidate, and he or she repays the loans via a lower initial salary. Similarly, while British Airways does not sponsor students as fully as it has in the past, when it does, the students pay for their training through their salary once they begin working for the airline. On the other hand, KLM airline does not sponsor candidates; however, it does partially fund an insurance policy to help cover the default risks for the banks that loan the money to the students for training as a result of early termination due to poor performance, failed medical examination, and other unforeseen circumstances. If the insurance policy is executed, students are contractually obligated to cease pursuit of a pilot career. U.S. airlines do not sponsor students for initial pilot training. As mentioned previously, pilot training in the United States is provided to individuals based on the self-sponsored concept—if they can pay, they can train. Students do not need to meet certain qualifications to train. However, once hired, all U.S. commercial airlines are required to provide the advanced training for their pilots. Several officials stated that most pilot students in the United States are not interested in becoming commercial airline pilots and pursue training to fly recreationally.

According to literature, most pilot schools in the United States conduct little screening of students that apply, whereas European schools routinely use comprehensive mechanisms to identify the most qualified students as to provide the best pilot candidates to their sponsoring airlines. The candidate screening process is generally the same across Europe and includes several interviews, various psychological tests, and scenario-based testing. According to some European officials, pilot schools in Europe focus on selection procedures and aptitude screening. According to officials at European airlines, basing selection on fixed standards instead of selecting candidates influenced by commercial pressure assures airlines that they are training a qualified pilot. Officials at these European airlines and the pilot schools they sponsor noted that pilot training in Europe is very expensive, and selecting the right pilot candidates is important because of safety reasons and the upfront investment for the company. In the United States, however, the most intensive screening process occurs when pilots seek employment with airlines. The airlines independently assess candidates' work experience, technical, and non-technical skills before hiring. (For more information on the demand for and supply of pilots, see app. II.) According to some aviation stakeholders, U.S. pilot schools would lose revenue if they screened or selected students as is done in Europe and would be challenged under privacy and anti-discrimination laws.

FAA Completed Most Required Inspections of Pilot Schools, but Oversight of Examiners and Instructors Is Difficult to Assess

The Majority of Required Inspections of Pilot Schools Were Completed from Fiscal Years 2006 through 2010

For fiscal year 2010, our analysis of FAA's PTRS data found that FAA completed about 78 percent of the required inspections for the 545 pilot schools with a Part 141 certificate (vocational schools and most collegiate aviation schools). As part of its oversight role, FAA monitors pilot schools with a Part 141 certificate to ensure that they meet the required safety and training regulations. To fulfill NPG requirements, FAA requires its inspectors to conduct on-site inspections of each of these schools at least once a year. The inspections focus on five areas related to pilot school operations and the airworthiness of training aircraft. Under operations, an inspection must cover the school's facility and student records. Under airworthiness, an inspection must cover the pilot school facility, compliance with FAA's airworthiness directives, and a Part 141 ramp check.[60] Inspectors enter the details and results of their monitoring in FAA's PTRS data system. In reviewing PTRS data for fiscal year 2010, we found that FAA completed the majority of the required inspections of schools with a Part 141 certificate—that is, the inspections covered all five inspection areas during the year. However, we found that for 118 of the 545 schools that were inspected during the fiscal year, all of the five inspection areas were not covered (see table 3). An annual inspection covering all of the inspection areas is important for overseeing these schools because it provides some assurance to FAA that they are meeting the regulatory requirements for providing adequate training to pilot students.

Table 3. Completion Percentage of Required Inspections for Pilot Schools with Part 141 Certificates for Fiscal Year 2010

		2010	
Required NPG inspection areas completed	Number	Percent	Cumulative percent
0 of 5	20	3.7	3.7
1 of 5	17	3.1	6.8
2 of 5	58	10.6	17.4
3 of 5	16	2.9	20.3
4 of 5	7	1.3	21.6
5 of 5	427	78.3	100.0
Total	545	100.0	

Source: GAO analysis of FAA data.

We were unable to determine whether the data were missing because they were entered incorrectly into PTRS, or because the inspections did not take place as required. We were also unable to determine the extent inspections were completed in fiscal years prior to 2010 due to limitations in tracking the number of pilot schools that existed in each fiscal year. FAA does not maintain a historical listing of the active pilot schools with a Part 141 certificate for a given fiscal year and, thus, we could not define the universe of pilot schools that was required to be inspected during fiscal years 2006 through 2009. As a result, we could not determine the identity and number of schools that needed to be inspected. While FAA officials recognized that all required NPG inspections are expected to be completed within each fiscal year time frame, they provided several reasons to potentially explain why some of the required inspections are not recorded in PTRS. FAA officials said that inspectors may have conducted the required inspections for some schools but incorrectly entered the details in PTRS.

For example, some FAA inspectors may conduct full inspections of schools that cover the five inspection areas, but may enter only two of the five inspection numbers into PTRS. The officials also said that inspectors had additional duties—such as following up on previously identified issues or addressing the need for additional oversight for certain inspection areas—in conducting inspections for some schools, which can make covering all five inspection areas difficult.

In addition, many of the part 141 pilot school inspections required by NPG are incorrectly recorded as discretionary inspections in PTRS making it difficult to use PTRS to determine if FAA had conducted all of the required inspections of part 141 pilot schools for a given year.

Specifically, 35 percent of the 4,551 part 141 pilot school inspections required by NPG in fiscal year 2010 were listed incorrectly in PTRS as discretionary inspections.

Also, 32 percent of the required inspections were listed as discretionary from fiscal year 2006 through 2010. As a result, those inspections would not show up in a list generated in PTRS of required inspections for fiscal year 2010. FAA officials said that this problem is likely caused by the program that populates PTRS with the NPG requirements.

When inspections or other sources revealed compliance issues or violations, FAA uses a variety of actions to enforce safety standards and regulatory compliance, such as oral or written counseling, administrative action, legal enforcement action, and referral for criminal prosecution. When an FAA inspector identifies a potential violation, he or she initiates an investigation and if FAA determines that a violation has occurred, FAA has a wide range of options available for addressing it. From fiscal years 2006 through 2010, our analysis of FAA's EIS data found that FAA initiated 230 enforcement cases against pilot schools with a Part 141 certificate. The majority of these cases resulted from an inspection of a school, but others may have resulted from other sources. During fiscal years 2006 through 2010, FAA used a wide range of enforcement actions against pilot schools with a Part 141 certificate (see Figure 7).

- *No action*: FAA can determine that *no action* is warranted. In 26 of the 230 cases (about 11 percent), enforcement cases were initiated, but no enforcement action was taken.
- *Administrative Actions*: In 186 cases (about 81 percent), FAA used *administrative actions* to address violations. These actions refer not only to warning notices and letters of correction but also informal actions, such as oral or written counseling, which can also be used by inspectors to address an apparent violation, provided that certain criteria are satisfied and the apparent violation is a low safety risk.
- *Enforcement actions*: We found that FAA rarely used punitive means such as revoking licenses and assessing penalties against pilot schools with a Part 141 certificate. FAA assessed monetary *civil penalties* in 12 cases (about 5 percent) for pilot schools with a Part 141 certificate and the sanctions ranged from \$500 to \$20,000. FAA *revoked the operating certificates* of schools in 3 cases, or slightly more than one percent. To illustrate the severity of an action that leads to revoking operating certificates, these include knowingly permitting school training aircraft to be used to carry illegal controlled substances or an intentional action to improperly credit training to or graduate students.

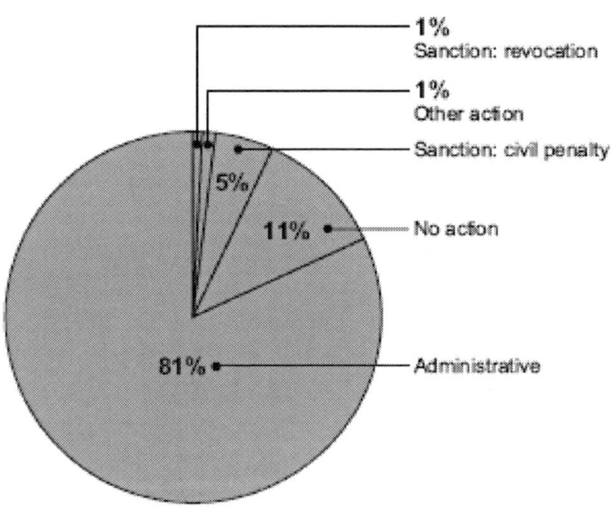

Source: GAO analysis of FAA data.
Note: Numbers do not add up to 100 percent due to rounding.

Figure 7. The 230 Enforcement Cases for Pilot Schools with Part 141 Certificates from Fiscal Years 2006 through 2010.

It Is Unclear Whether Required Inspections for Pilot Examiners Were Completed Because FAA Data System Lack Historical Information

From fiscal years 2006 through 2010, our analysis of FAA's PTRS data found that FAA completed 9,016 inspections of pilot examiners, but it is unclear whether FAA met all of its oversight requirements in this area (see Figure 8). FAA uses private individuals or organizations to supplement its workforce and to provide certification activities such as examining and testing of pilot applicants for a fee paid for by the applicant.[61] Known as designees, pilot examiners are generally appointed by FAA's local district personnel for either 3 years (for an individual) or 5 years (for an organization).[62] As part of its oversight role, FAA requires each pilot examiner to be inspected by FAA inspectors at least once annually and high activity pilot examiners must be inspected at least twice annually, as outlined in the agency's oversight policy and NPG directives.[63] Additionally, several other circumstances may require an FAA inspector to inspect a pilot examiner, such as noncompliance with the applicable certification policies, an excessively high certification passing rate, or involvement in an accident, incident, or other violation.[64] Although we know the number of inspections conducted for each fiscal year, we could not determine the completion percentage of the required inspections for either the routine annual or additional inspections for high activity or special circumstances for each fiscal year due to limitations in available data for the population of pilot examiners.

Although we could not determine the completion percentage of the required inspections for pilot examiners, our analysis of PTRS inspection data showed 1- and 2-year gaps in the oversight of some pilot examiners. For instance, we found 114 pilot examiners with a 1-year gap between inspections, and 11 pilot examiners with a gap of 2 years. This may indicate that required inspections of pilot examiners were not completed by FAA in a given fiscal year or that inspections were unnecessary due to inactivity of the examiners during that year. FAA officials told us that, until recently, FAA had not analyzed the extent to which it has conducted all required pilot examiner inspections on a national level.

FAA has previously taken steps to improve oversight of pilot examiners, but still faces issues in this area. In 2005, FAA developed 14 recommendations to improve pilot examiner compliance, 11 of which were implemented. FAA officials told us that the implementation of these recommendations has resulted in improvements in pilot examiner oversight guidance and in the information technology used to oversee pilot examiners for local district offices. They also said that, as a direct result of the recommendations, more of the pilot examiners with poor performance issues have been terminated and oversight of pilot examiners has improved.

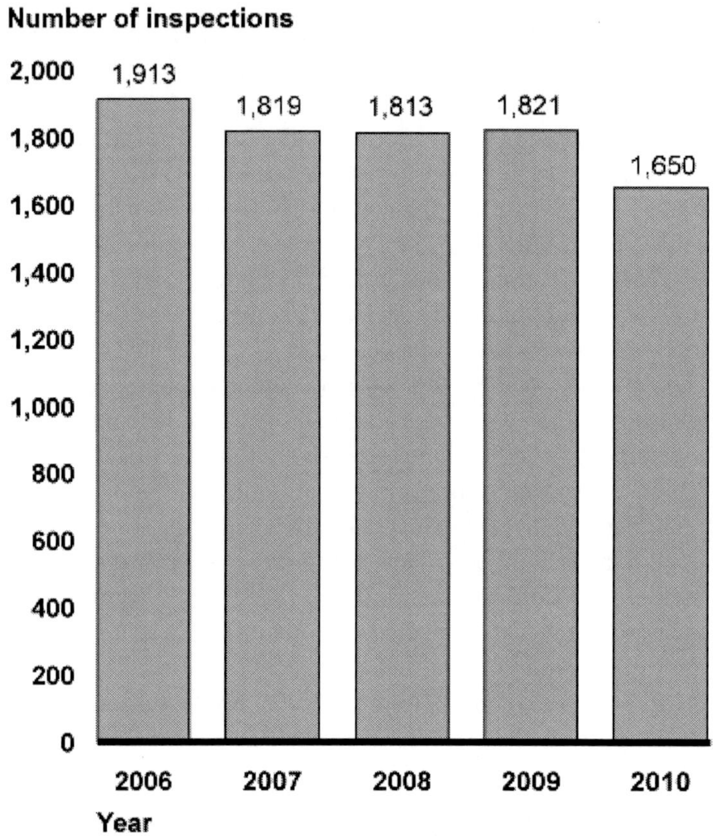

Source: GAO analysis of FAA data.

Figure 8. The 9,016 Inspections of Pilot Examiners from Fiscal Years 2006 through 2010.

Nevertheless, FAA officials acknowledged they still face some issues in oversight of pilot examiners, due, in part, to the FAA's current data systems' difficulty compiling inspection data at the regional or national level. In September 2010, FAA began developing quarterly assessment reports covering 12-month periods on the oversight of its designees, including pilot examiners, to assist in identifying oversight gaps and potential areas of concerns. We reviewed the quarterly reports that covered July 2009 through March 2011 and found they identified a number of areas of concern regarding the oversight of pilot examiners.

For example, some pilot examiners with the highest activity had not been inspected over the previous 12 months. FAA began creating the quarterly assessment reports to better inform management officials at the national office level.

For example, in the most recent report provided by FAA officials, five high-activity pilot examiners were identified that performed a total of 1,623 pilot practical tests, but no inspections were conducted for these pilot examiners for the previous 12 months. Conversely, the report found that FAA conducted 218 inspections of the 171 pilot examiners with the lowest testing activity during the same period.[65]

Based on this report, FAA's inspection record is not consistent with the pilot examiners that are responsible for conducting the largest numbers of practical tests, and conducting more oversight on the examiners conducting significantly fewer tests.

FAA officials said that the quarterly assessment reports are a temporary way of assessing the extent to which it is conducting all required inspections, but FAA is in the process of developing a new designee management system that it expects to be operational by July 2012. FAA officials told us that the new system is being designed to provide more comprehensive data on designees, including pilot examiners, by combining data that FAA currently maintains in various data systems. Thus, the new system will consolidate the management and oversight functions for designees to allow for more readily available data. The officials also told us that they are in the process of revising oversight policy for designees and improving tools for selecting which pilot examiners to inspect.

While Not Required, FAA Conducts Optional Oversight Inspections for Flight Instructors, but to What Extent Is Unclear

Unlike oversight of pilot schools with Part 141 certificates and pilot examiners, annual inspections of individual flight instructors (i.e., under Part 61 regulations) are not required by FAA. From fiscal years 2006 through 2010, our analysis of FAA's PTRS data found that FAA completed 1,761 inspections of flight instructors. Oversight for flight instructors is generally limited to initial and subsequent certification renewal, but additional oversight of flight instructors is conducted as an optional work activity by FAA. However, despite being optional, FAA officials from one local district office told us that their inspectors make this area a priority in their planned work activities for a given fiscal year.

According to FAA policy, flight instructor certificates are renewed every 24 months. However, inspections of flight instructors and their training activities should take place on a random basis in the interim, but should be prioritized as a result of, for instance, observations of noncompliance actions made during a pilot school inspection, an instructor or student is involved in an accident or incident, or when an instructor has a student failure rate of 30 percent or greater on FAA's certification tests. The oversight of flight instructors is important because, like the examiners, this population serves as a gatekeeper for ensuring that pilot students are being properly trained as they seek certification. However, similar to pilot examiners, we could not determine the extent of oversight that FAA provided for the entire flight instructor population or the specific reasons that inspections were conducted during the 5 fiscal years.

Similar to pilot schools, when inspections or other sources revealed compliance issues or violations, FAA uses a variety of actions to enforce safety standards and regulatory compliance. For fiscal years 2006 through 2010, our analysis of FAA's EIS data found 178 cases against flight instructors (see Figure 9).

- No action: FAA can determine that no action is warranted. In 38 of the 178 cases (about 21 percent), enforcement cases were initiated but no enforcement action was taken.
- Administrative Actions: In 109 cases (about 61 percent), FAA used administrative actions to address violations. These actions refer not only to warning notices and letters of correction but also informal actions, such as oral or written counseling, which can also be used by inspectors to address an apparent violation.
- Enforcement actions: We found that FAA rarely used punitive means such as suspending or revoking licenses and assessing penalties against flight instructors. FAA suspended licenses in 9 cases (about 5 percent) and revoked licenses in 16 cases (about 9 percent). FAA also assessed monetary civil penalties in three cases (about 2 percent) for flight instructors.

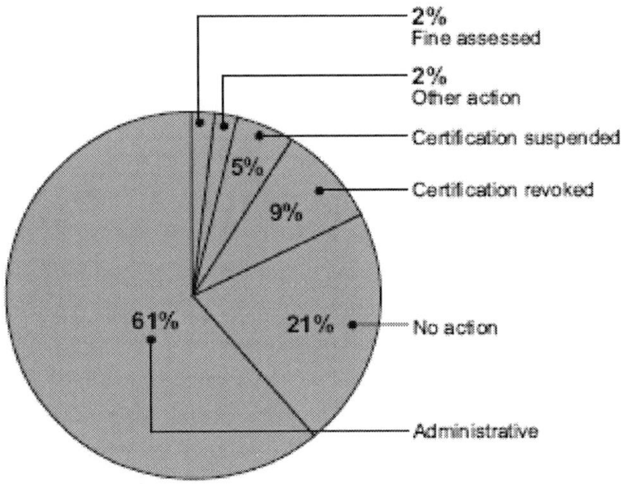

Source: GAO analysis of FAA data.

Figure 9. The 178 Enforcement Cases for Flight Instructors from Fiscal Years 2006 through 2010.

Inspectors also Use the Safety Performance Analysis System to Plan Their Work

In addition to using PTRS and EIS, FAA inspectors develop their work plan using another somewhat limited oversight tool—the Safety Performance Analysis System (SPAS)—which provides data access and analysis on pilot schools, pilot examiners, and flight instructors.[66] The SPAS system is a data analysis application that monitors performance measures and calls attention to trends. SPAS builds on inspection results and other data and is intended to assist FAA's local district offices in applying their limited inspection resources to those entities and areas that pose the greatest risk to aviation safety. For example, when particular inspection tasks warrant attention, SPAS will "flag" an advisory notification to a FAA inspector and prompt the inspector to look into the situation, e.g., a flight instructor with a high rate of student failure on practical tests. While SPAS may be useful at the local level, it does not have the capability to perform national level rollup analysis. Thus, FAA does not have the ability to measure FAA's annual performance in meeting the NPG inspection requirements for pilot schools and pilot examiners based on PTRS inspection data, or to make risk-based, data-driven decisions about the scope of its discretionary, planned work activities that include flight instructors. As a result, FAA's national office cannot readily access comprehensive inspection completion data and determine the level of oversight its field staff is providing for pilot schools and pilot examiners.

CONCLUSION

Public and media concerns about aviation safety escalated as a result of the Colgan crash in early 2009, and Congress and FAA have taken steps to improve aviation safety by making revisions to the training requirements that airlines must provide for commercial airline pilots. Our analysis indicates that FAA has an opportunity to ensure that the initial pilot training process for producing pilots' commercial certificates is still relevant for the necessary knowledge and skills for airline positions. Although many of the improvements to training being considered are for existing airline pilots—such as those related to decision making and operating in a crew environment—they may also apply to initial pilot training. We are not making recommendations in this area, because FAA has initiated some efforts and has plans for other efforts to address pilot training issues. For example, FAA plans to establish a government and industry working group during fiscal year 2012 to address issues related to pilot certification testing standards and training. Therefore, we encourage FAA to continue its efforts, with industry and academia collaboration, in reviewing the initial pilot training process, including ground school content, training hour requirements, and knowledge testing for commercial pilot certification under the Part 61 regulations.

FAA's oversight of key functions for the initial pilot training process in the United States by which commercial pilots become certificated is reasonably sound. We found that FAA completed most required inspections of vocational pilot schools (Part 141), but the agency's data sources did not provide certainty that the level of oversight is consistently being performed in accordance with the agency's guidelines and policies, including oversight requirements for pilot examiners. FAA's national office does not monitor the annual completion of the requirements outlined in the annual NPGs related to monitoring pilot schools and pilot examiners. Therefore, the national office cannot readily determine the level of oversight its field staff is providing for these key stakeholders and has been unable to produce this information. Better internal control mechanisms, such as creating and reviewing agencywide reports using PTRS data periodically, would improve FAA oversight by providing assurance that all required inspections were completed. Additionally, FAA could establish standard procedures for entering PTRS data for inspectors at the local level of the completed inspection areas to avoid the uncertainty of whether or not they were completed as required. This would not only help FAA better measure its performance in meeting the annual NPG inspection requirements, but would also enable the agency to make more informed decisions about the scope of its discretionary, planned work activities for flight instructors. Further, the steps taken by FAA to develop a quarterly assessment report on the oversight and performance of pilot examiners is promising, but FAA should also consider developing a similar process for oversight of flight instructors because it could identify potential areas of concerns.

Recommendations for Executive Action

We are making two recommendations to improve FAA's oversight of pilot certification and training. The Secretary of Transportation should direct the Administrator of the Federal Aviation Administration to develop a comprehensive system that may include modifying or improving existing data systems to

- measure performance for meeting the annual National Program Guidelines' inspection requirements for pilot schools with a Part 141 certificate and pilot examiners and
- better understand the nature and scope of the discretionary, planned inspections for flight instructors.

Agency Comments and Our Evaluation

We provided a draft of this report to DOT for review and comment. In responding to our recommendations, FAA officials said that they agreed that improvements in oversight data were needed but indicated that the quarterly assessment reports already measure the level of oversight of pilot examiners and summary data is being provided to the national office and regional division managers. Additionally, the designee management system currently under development will address the recommendation for pilot examiner designees in a more permanent way. We retained the recommendation; our report notes the oversight improvements underway for pilot examiners, and we will assess the effectiveness of the designee management system once it is implemented. FAA also provided technical comments that we incorporated as appropriate.

Gerald L. Dillingham, Ph.D.
Director, Physical Infrastructure Issues

APPENDIX I: SCOPE AND METHODOLOGY

To address our objectives and the pilot supply appendix, we reviewed and synthesized literature on pilot training and certification in the United States. Specifically, we reviewed a range of reports from GAO, Federal Aviation Administration (FAA), Congressional Research Service, International Civil Aviation Organization, National Transportation Safety Board, and Bureau of Labor Statistics that included general background information on a variety of related issues on U.S. pilot training, such as pilot certification and training issues in the U. S.; historical trends, current supply and demand, and forecasts for commercial airline pilots; types and requirements of pilot training schools; FAA regulatory training requirements for different levels of pilot certification; FAA oversight of U.S. pilot training system; and comparable international pilot training systems. We conducted a literature review search from databases, such as ProQuest, TRIS, and Lexis, as well as well as trade publications and literature from aviation stakeholders. Furthermore, we reviewed the Federal Aviation Regulations related to training and certification for pilots under Part 61, Part 141, and Part 142. We also reviewed FAA regulations, policy, and oversight documentation related to pilot training, management guidance for overseeing pilot schools, designated pilot examiners (pilot examiner), and certified flight instructors (flight instructor). Additionally, we reviewed FAA's national program guidelines for fiscal years 2006 through 2010 regarding the required and planned oversight activities as well as FAA's practical test standards used to certificate pilots as private, commercial, and airline transport. We also reviewed provisions of the Airline Safety and Federal Aviation Administration Extension Act of 2010 (Pub. L. No. 111-216) related to "Flight Crewmember Screening and Qualifications" and "Airline Transport Pilot Certification."

To review FAA's inspection and enforcement activities related to pilot schools, pilot examiners, and flight instructors, we obtained FAA's inspection and enforcement policies and analyzed raw data from FAA's inspection and enforcement databases. We analyzed data from the Program Tracking Reporting System (PTRS) for inspections that closed (had a closing date) in fiscal years 2006 through 2010 and data from the Enforcement Information System (EIS) for enforcement actions with a date of final action in those fiscal years. To assess the reliability of the inspection and enforcement data that we received from FAA, we performed electronic testing of the data elements that we used, obtained and reviewed documentation about the data and the systems that produced them, and interviewed knowledgeable FAA officials. We used these data to determine the extent to which FAA had completed all required inspections of pilot schools and pilot examiners, and planned inspections of flight instructors. For pilot schools with a Part 141 certificate, we analyzed data from PTRS on the numbers of required inspections that FAA completed, and whether all five inspection areas were covered, and compared to the requirements set in FAA's National Program Guidelines (NPG). The NPG for each fiscal year indicated that an inspection was required for each school within each region and to include under operations: (1) air agency facility inspection (PTRS activity number 1640) and (2) student records (PTRS activity number 1649). The NPG for each fiscal year also indicated that an inspection was required for each school within each region and to include under airworthiness: (1) pilot school facility (PTRS activity number 3650), (2) airworthiness directive compliance (PTRS activity number 3667 or 5667), and (3) part 141 ramp (PTRS activity number 3664 or 5664). To determine the nature and scope of the enforcement actions that FAA closed against pilot schools with a Part 141 certificate, we analyzed data on these actions from EIS, including whether the actions were administrative, fines, or suspensions or revocations of schools' Part 141 operating certificates. We also analyzed data to determine the minimum, median, and maximum dollar amounts of fines and durations of suspensions. We tested the reliability of the PTRS and EIS data that we received from FAA with electronic testing of the data elements that we used, obtained and reviewed documentation about the data and the systems that produced them, and interviewed knowledgeable FAA officials. We found the data to be sufficiently reliable for our purposes.

For pilot examiners, we analyzed data from PTRS on the numbers of required inspections that FAA completed and compared to the requirements set in the NPG. The NPG for each fiscal year indicated that an inspection was required for each pilot examiner within each region under: pilot examiner—large-turbojet (PTRS activity number 1664) and pilot examiner—other (PTRS activity number 1665). We also obtained and reviewed summary data from FAA contained in quarterly assessment reports on the oversight of its designated representatives, including pilot examiners. The quarterly reports contained data from July 2009 through March 2011.

For flight instructors, the PTRS activity number 1662 is used to record CFI inspections in PTRS. The NPG for fiscal years 2009 through 2010 did not indicate inspections for certificated flight instructors as a planned work activity. To determine the nature and scope of the enforcement actions that FAA closed against flight instructors, we analyzed data on these actions from EIS, including whether the actions were administrative, fines, or suspensions or revocations of flight instructors' certificates. We also analyzed data to determine the minimum, median, and maximum dollar amounts of fines and durations of suspensions.

We interviewed government officials at the Federal Aviation Administration and National Transportation Safety Board. We conducted semistructured interviews with representatives from a range of aviation stakeholder organizations (see below). We also interviewed researchers involved in the pilot source study. We visited pilot training stakeholders in six states included in FAA regions that had higher number of pilot schools, higher number of pilot certificates issued in 2009, while taking into consideration the presence of FAA regional and district offices and regional airlines' offices in some locations. Thus, in our visits to Arizona, Florida, Georgia, Indiana, Maryland, and Utah, we interviewed officials at FAA regional and district offices, regional airlines, pilot schools, and college aviation schools. However, because we selected these six states as part of a nonprobability sample, our findings cannot be generalized to all pilot training stakeholders in the United States. Through the combination of site visits and semistructured telephone interviews, we interviewed representatives of 24 regional airlines that transported about 97 percent of regional passengers in 2009, according to the Regional Airlines Association's 2010 annual report. (See table 5.)

Table 4. Agencies, Organizations, Associations, Schools, and Airline Interviewed

U.S. federal agencies
Department of Transportation Office of Inspector General
Federal Aviation Administration
FAA Civil Aerospace Medical Institute
National Transportation Safety Board
International organizations
International Civil Aviation Organization
International Air Transport Association

U.S. federal agencies
Industry associations
Air Line Pilots Association
Air Transport Association
Aviation Accreditation Board International
Coalition of Airline Pilots Association
Flight Safety Foundation
Flight School Association of North America
National Association of Flight Instructors
Professional Aviation Board of Certification
Professional Aviation Safety Specialists
Regional Airline Association
Society of Aviation and Flight Educators
University Aviation Association
Industry organizations
Airline Apps, Incorporated
FltOps.com
Pearson Education, Incorporated
SIM Services
Pilot schools
Aerosim Flight Academy
Airline Transport Professionals
Arizona State University College of Technology and Innovation
Aviation Flight Academy, Incorporated
Dowling College School of Aviation
Embry-Riddle Aeronautical University
Phoenix East Aviation
Purdue University Department of Aviation Technology
University of North Dakota Department of Aviation
Airlines
Air Wisconsin Airline Corporation
American Eagle
Atlantic Southeast Airlines
Comair
CommutAir
Delta Air Lines
ExpressJet
Horizon Air
Kenmore Air Harbor
Mesa Air Group, Inc. (Mesa Airlines and Freedom Airlines)
Piedmont Airlines
Pinnacle Airlines Corp. (Pinnacle Airlines, Colgan Air, and Mesaba Airlines)
PSA Airlines

Table 4. (Continued)

U.S. federal agencies
Republic Airways Holdings (Republic Airlines, Chautauqua Airlines, Frontier Airlines, Shuttle America Corp.)
SkyWest Airlines
Trans States Holdings, Inc. (Trans States Airlines, GoJet Airlines, Compass Airlines)

Source: GAO.

Table 5. Organizations Contacted by Country

France
Directorate General of Civil Aviation
Ecole de Pilotage Amaury de la Grange
Air France Airlines
Corsairfly (regional airline)
International Civil Aviation Organization (ICAO) European Regional Office
Aircraft Owners and Pilots Association (AOPA) France
Germany
European Aviation Safety Agency (EASA)
Lufthansa Flight Training Pilot School
Netherlands
Ministry of Transport, Public Works and Water Management
Joint Aviation Authorities Training Organization (JAA TO)
KLM Airlines
KLM Flight Academy
Nationale Luchtvaart School (training organization)
United Kingdom
United Kingdom Civil Aviation Authority
Oxford Aviation Academy
Cabair College of Air Training (Cranfield)
British Airways
Virgin Atlantic
Willowair (training organization)
Aircraft Owners and Pilots Association United Kingdom
General Aviation Safety Council
Royal Aeronautical Society
Flight Safety Committee
Peter Metcalfe (flight instructor)

Source: GAO.

In addition, we conducted site visits to the France, Germany, Netherlands, and United Kingdom. We focused on these European countries' pilot certification and training requirements because they offer a different model than the United States. The site visits allowed us to obtain information on European countries' pilot standards, as well as their efforts to revise their training requirements from traditional training objectives and methodology to competency-based training models. However, because we selected these four countries as part of a nonprobability sample, our findings cannot be generalized to all European countries. During these site visits, we interviewed officials at the European Aviation Safety Agency (Europe's aviation regulator), civil aviation authority officials, representatives from international and European aviation stakeholder groups, representatives from airlines, and flight training organizations (schools). However, because these four countries were selected as part of a nonprobability sample, the findings from our interviews cannot be generalized to all European countries. (See table 6.)

We conducted this performance audit from March 2010 through November 2011 in accordance with generally accepted government auditing standards. Those standards require that we plan and perform the audit to obtain sufficient, appropriate evidence to provide a reasonable basis for our findings and conclusions based on our audit objectives. We believe that the evidence obtained provides a reasonable basis for our findings and conclusions based on our audit objectives.

APPENDIX II: NATURE AND SCOPE OF THE SUPPLY OF AND DEMAND FOR AIRLINE PILOTS

For several years now, the issue of whether the United States will maintain an adequate supply of qualified pilots has emerged in literature, media sources, and aviation industry discussions, but the scope of the supply of and demand for U.S. airline pilots is unknown and difficult to determine. The number of pilot certificates held in the United States has been declining and is expected to continue to decline in the near future. Demand, which has historically fluctuated, is projected to significantly increase over the next 20 years. Nevertheless, the International Civil Aviation Organization (ICAO) is predicting a surplus of pilots in North America, based, in part, on the regional capacity to provide pilot training, level of current and projected pilot population, and the slow growth rate of the aircraft fleet requiring additional pilots to operate them. However, certain factors could affect that projected surplus, including the 2007 legislation extending the mandatory retirement age for airline pilots and the sharp curtailment of military pilots as a hiring resource. Particularly given the decrease in the number of military pilots available to be hired by airlines, the number of students who enroll in and complete training in pilot schools will also affect the supply of commercial pilots, as the availability of pilots for entry-level regional airline jobs can be directly linked to the number of pilot certificates issued. Student enrollment in pilot schools is purported to be declining and the dropout rate for completing pilot training is high. Furthermore, others factors present potential challenges to the pilot training industry, such as available financial sources to fund pilot training and the impact of the 2010 legislation requiring additional pilot certification requirements for airline first officers.

The Economy and Demographics Affect Airline Pilot Supply and Demand

The safety and economic contribution of the air transportation system not only relies on having well-trained airline pilots but also on having enough of them to meet demand. The demand for commercial pilots is a function of the size the commercial airline fleet and the number of pilots needed to operate that fleet size. Demand for professional pilots in the United States has historically fluctuated, driven by a number of factors including consumer demand for air travel and the general state of the economy, fuel prices, regulatory changes, and aircraft fleet changes. For example, since 2008, the economic recession has significantly affected the U.S. airline industry. Several airlines have filed for bankruptcy and ceased operations, while other airlines have merged. It is still unknown when and to what extent the U.S. airline industry will fully recover, but demand for air travel is highly cyclical and largely reflective of the state of the economy, and expected to increase significantly over the next 20 years. The total number of pilot certificates held in the United States has been fluctuating downward over the last 10 years. The number of active student, private, commercial, and airline transport pilot certificates held in the United States decreased from 608,079 in 2000 to 554,237 in 2009, or 9 percent. The Aircraft Owners and Pilots Association (AOPA) predicts that, by 2014, the number of pilots in the United States will decline to 500,000.[1]

In late 2010, Boeing projected the commercial aviation industry in North America will require over 97,000 new pilots over the next 20 years (over 466,000 new pilots globally) to accommodate the strong demand for new and replacement aircraft, but noted that emerging markets in Asia, especially in China, will experience the greatest need for pilots. As demand for air travel continues to increase internationally, the demand for qualified pilots will likely continue to grow in many parts of the world, particularly in Asia and the Middle East and possibly attract qualified commercial pilots from the U.S. market. For instance, Boeing forecasts that China will need over 70,000 new pilots by 2029 for new commercial aircraft. In order to deal with the lack of available pilots today, some foreign airlines have started training their pilots through established flight training organizations or sending their students to pilot schools in English speaking countries, such as the United States. However, if industry projections are realized and commercial airlines, both mainline and regional, need large numbers of pilots, it is uncertain whether an adequate supply of qualified pilots will be available.

However, despite these projections, ICAO is reporting a likely surplus of pilots in North America. ICAO noted in its 2011 report on the 20-year global and regional forecast on civil aviation personnel that the United States, with close to 270,000 professional pilots (i.e., commercial or airline transport), accounts for approximately 58 percent of the global pilot population. Furthermore, the United States has an annual training capacity for producing about 27,000 professional pilots annually. Based on the projected annual need for new pilots and considering several factors, ICAO determined that the United States will likely have a surplus of pilots under all of the scenarios it considered for the North American region (i.e., the United States and Canada). This may seem surprising considering the general perception that a pilot shortage in the United States is inevitable, but was supported by ICAO based on the following reasons. First, the annual training capacity in the United States represents 60 percent of the worldwide training capacity; thus, the ICAO calculations indicate that North America has more than enough training capacity to train the number of new pilots needed in this region. Second, fleet growth rates—i.e., the number of aircraft needing pilots—for the North American region for which the United States is the dominant player, are relatively low. Third, the current number of licensed professional pilots in the United States represents 58 percent of the global pilot population. Although these factors do point to a surplus, other economic and demographic factors that can be difficult to determine could affect the projected surplus.

The temporary impact of recent legislation could affect the availability of commercial pilots in the future and other measures may be required to address related long-term issues.

When recent concerns arose about a potential pilot shortage, the Fair Treatment for Experienced Pilots Act was enacted (age 65 law) in 2007 and allowed pilots to fly domestic routes until age 65 instead of the prior mandatory retirement age of 60.[2]

For international flights, one pilot may be up to age 65 provided the other pilot is under age 60, consistent with the November 2006 ICAO standard. This effort attempted to ensure there would not be a large group of highly skilled pilots retiring at the same time, but according to ICAO data, pending age-related retirements will begin to reemerge in 2012.

The 2011 ICAO report also indicated that approximately 85,000 professional pilots (31 percent of the total pilot population) are aged 55 or more; among this age group, 34,000 (13 percent of the total pilot population) will be eligible for retirement by 2015. Another 34,000 will be eligible for retirement by 2020. The report also asserted that it is likely, considering the age structure of pilots in North America, that retirements will take place at an accelerating rate, thereby contributing to a potential increase in future pilot demand.

However, all of the pilots holding commercial or airline transport pilot certificates in the United States may not be automatically hired by airlines due to performance and skills deficiencies. Therefore, despite a projected surplus of professional pilots, having only a potentially limited supply of pilots that airlines determine to be qualified to hire could still affect the air transportation industry in the United States.

In addition, one of the traditional, preferred sources of airline pilots has been severely curtailed, which will likely continue to affect the availability of pilots in the future. U.S. commercial airlines are no longer hiring as many pilots from the U.S. military as they had in the past, which has historically provided airlines with a steady supply of highly qualified pilots. According to literature sources, until the 1990s, roughly 90 percent of the pilots hired by mainline airlines came from the U.S. military, with the remaining source being from civilian aviation.

In 2008, about 30 percent of pilots hired by commercial airlines had military training backgrounds. Moreover, the military is training fewer pilots and retaining more of them with better pay and other financial incentives. In addition, many commercial airline pilots from Vietnam era are reaching retirement age.

Domestic Student Enrollments in U.S. Pilot School Is Declining and the Dropout Rate of Student Pilots Is High

Particularly given the decrease in military pilots as a source of potential civilian pilots, the number of students who enroll in and graduate from pilot schools also affects the supply of pilots, and could affect the ICAO projected surplus of pilots for the North America region, specifically in the United States. The availability of pilots for entry-level regional airline jobs can be directly linked to the number of student pilot starts and the number of pilot certificates issued. In the past few years, the Federal Aviation Administration (FAA) has revised its forecast to reflect an uncertain number of student pilots. In 2008, FAA predicted 100,200 student pilots in the year 2025, but lowered that estimate in its 2009 through 2025 forecast report to 86,600. However, in FAA's 2011 through 2031 forecast report, FAA predicted the number of student pilots increasing to 113,500 in 2025 (120,600 in 2031), due to a rule change by FAA that makes student pilot certificates valid longer for pilots under the age of 40.[3] Also, from 1999 through 2009, the number of private certificates issued decreased about 26 percent, but commercial certificates slightly increased about 10 percent. Representatives of the pilot training industry that we interviewed told us that the overall trend reflects a lower number of domestic students starting pilot training, a trend which continued into 2011. Representatives from a few pilot schools told us that pilot schools across the United States experienced large declines in enrollments in 2010. For example, the University of Illinois decided to eliminate its 65-year-old aviation school after a decade of declining enrollment. The collegiate aviation school produced fewer degrees and served fewer students than any other program on the campus during the previous school year and no aviation students had been accepted for the coming semester.

The decline of student pilot certificates and student dropout rate are concerns to AOPA. According to AOPA, the number of current student pilot certificates is the best indicator of the supply of future pilots. Representatives of several pilot schools with whom we spoke told us that they were becoming more and more dependent on foreign students to maintain their pilot training operations. Also, according to FAA officials, representatives of AOPA, four pilot schools, and several European organizations we interviewed said that some European and Asian airlines and pilot training organizations send students to the United States for pilot training to take advantage of the relatively inexpensive fuel and the year-round weather conditions for training in states such as Florida and Arizona. In addition, the dropout rate of student pilots could affect the supply of pilots. According to a 2010 AOPA study, almost 80 percent of student pilots drop out of training for four key reasons: (1) lack of educational quality, (2) lack of customer focus (not a good value or pricing not competitive), (3) insufficient sense of community (lacks an atmosphere that makes students feel welcome in the aviation community), and (4) lack of information sharing (school does not provide realistic estimate of time and costs required for a pilot certificate and statistics on student success rates at schools).[4] The study indicated that while cost of training was a factor, it was less significant than the other four reasons people did not complete training—even though some literature suggests otherwise as discussed below.

AOPA, as a result of this information regarding the dropout rate, has initiated efforts with close coordination with the pilot training industry to work on solutions to stop the outflow of students and to increase the pilot population. AOPA has begun an initiative focused on student retention in pilot training that will consist of a series of regional meetings across the United States to collect perspectives and industry input on potential improvements that can be made in providing pilot training. AOPA has met with representatives from the aviation community, including pilots, student pilots, aviation businesses, pilot schools, and flight instructors who are currently involved in conducting pilot training. In addition, the association has launched a Let's Go Flying Web site (www.aopa.org/letsgoflying) and created various publications to inspire more people to become interested in learning to fly by providing an introduction to and information about becoming a pilot.

Challenges to Financing Pilot Training Could Affect Enrollment

Aviation stakeholders from 4 of the 9 pilot schools, two industry associations, and one industry organization we interviewed told us that one of most important challenges for maintaining an adequate supply of students for pilot schools is the availability of financial support for pilot training. Pilot training costs vary amongst the pilot schools. According to AOPA and other sources, the cost of training from the beginning of training through a commercial pilot certificate and a multiengine rating could be in excess of $40,000 but varies across the country. According to AOPA, pilot training can cost about $100,000 or more at most collegiate aviation schools for a 4-year degree and the flight training provided. The University Aviation Association (UAA) indicated that there is no single comprehensive information source regarding the cost of pilot training across the United States, but AOPA officials told us costs vary between $40,000 and $100,000 for the training needed to obtain a commercial pilot certificate and a multiengine rating. Furthermore, as previously stated, pilots would likely be responsible for any increased costs if additional training requirements were developed by FAA. The costs are high compared to the low starting salaries. According to a 2009 survey by the Regional Airline Association (RAA) of member airlines based on 2008 salary data, the salaries for regional pilots generally averaged between $28,000 and $43,000 for regional airline first officers and $62,000 to $102,000 for regional airline captains. RAA estimated that salaries have increased 2 to 5 percent since but it does not have more current data. Some stakeholders we interviewed said that these realities could potentially prevent prospective students from enrolling in a pilot school or reduce the desire to become a pilot. However, it is important to note that the previously mentioned AOPA study reported that that training costs were not a statistically significant reason individuals dropped out of training.

A range of tuition resources is available, but some of these sources are drying up. To pay for pilot training, AOPA officials said that students typically use personal funds, personal credit (credit cards and personal loans), scholarships, grants, parent or student loans, other educational loans, and the Veterans Administration's (VA) Montgomery GI Bill benefits. For example, AOPA offers loans for pilot training up to $25,000 and allows flexible funding options for students since AOPA does not limit the use of funds to certain types of schools or training. One benefit of collegiate schools and some vocational schools is that they are generally eligible to receive VA benefits and federal financial aid (such as federal education loans or grants). However, flight instructor-based schools do not generally qualify for federal financial aid or VA benefits. VA benefits allow a qualified student to be reimbursed up to 60 percent for ground and pilot training costs to the maximum allowable limit, but do not fund training for a private pilot certificate. Many financial institutions have provided financing for pilot training through educational loan programs, such as the Federal Family Education Loan program, and personal loans.[5] According to literature we reviewed and representatives of four pilot schools and three industry associations we interviewed indicated that many of the private banks have been tightening restrictions on financing available to potential new pilot students, and others have left the pilot training loan market. The National Association of Flight Instructors reported that Sallie Mae took a $1 billion loss in 2009 for educational loans, which explains, in part, the reason it has become increasingly difficult to obtain funding.[6] We reported in November 2009 that many lenders offering student loans have exited the market in response to limited access to capital resulting from the credit crisis.[7] Thus, lenders have begun to give pilot training a higher risk profile than in the past and have been slowly exiting this loan market.

Recent Efforts to Improve Airline Safety and Pilot Training Could Impact Pilot Availability

Representatives of the regional airlines we interviewed and stakeholder associations have voiced a significant level of concern regarding the legislative mandate to increase the number of flight hours needed for first officers to be hired by commercial airlines. As discussed earlier in the report, the Airline Safety and Federal Aviation Administration Extension Act requires that FAA develop regulations requiring all airline pilots, including the first officers, to hold an airline transport pilot certificate—the highest level of pilot certification and requiring the highest number of total flight hours—instead of the commercial pilot certificate that is required today.[8] Aviation stakeholders have voiced significant concerns that requiring first officers for regional airlines to possess an airline transport pilot certificate will likely result in the inability to fill some positions due to the lack of qualified pilots.

Table 6. Airline Transport Pilot Certifications Issued from 1999 through 2009 from Pilot Schools Operating under Part 61, Part 141, and Part 142

Year	Part 61	Part 141	Part 142	Total
1999	6,666	4	503	7,173
2000	6,652	5	1,039	7,696
2001	5,735	2	957	6,694
2002	3,670	3	904	4,577
2003	3,087	10	844	3,941
2004	3,490	2	987	4,479
2005	3,579	6	1,130	4,715
2006	3,542	4	1,191	4,737
2007	4,873	4	1,123	6,000
2008	3,845	6	1,022	4,873
2009	2,078	2	815	2,895

Source: FAA.

Note: Training for the airline transport pilot certificate is generally provided by pilot schools that train students under Part 61 and Part 142 regulations.

The overall decreasing trends in pilot certificates being issued is illustrated in table 4 with the general decline in the number of airline transport pilot certifications issued from 1999 through 2009 from pilot schools operating under Part 61, Part 141, and Part 142 regulations. The number of airline transport certificates issued decreased about 60 percent.

Aviation stakeholders such as AOPA and RAA have both also voiced concerns that increasing pilot certification requirements to become an airline pilot could significantly discourage potential pilots from entering aviation due to the increased time and expense required to meet the new hiring minimums of airlines. For example, regional airline officials we interviewed said the new requirement will create a gap in experience for new pilots seeking entry level airline jobs that could take several years to fill. Furthermore, AOPA said the cost of obtaining the additional 1,250 flight hours needed to meet the total time for an airline transport pilot certificate could deter many new pilots into pursuing other professional careers by increasing the number of flight hours to obtain employment as an airline pilot. Additionally, the 2011 ICAO report mentioned earlier stated that even though the United States has sufficient training capacities, the requirement that all commercial airlines' first officers have an airline transport pilot certificate could drastically limit the availability of first officers to support existing delivery schedules for new aircraft at many airlines.

The majority of officials representing the regional airlines we interviewed also indicated that the proposed FAA rules to revise pilot duty and rest requirements will potentially create a need for more pilots.[9] Provisions in the Airline Safety and Federal Aviation Administration Extension Act of 2010 directed FAA to issue a regulation to specify limitations on the hours of pilot flight and duty time to address problems relating to pilot fatigue.[10] NTSB has long been concerned about the possible safety effects of fatigue in the aviation industry—specifically pilots—and have issued several safety recommendations to FAA after identifying it as a contributing factor in several aviation accidents. FAA identified the issue of pilot fatigue as a top priority following the Colgan Air crash, and the NTSB accident report stated the pilots' performance was likely impaired because of fatigue, but the degree to which it contributed could not be conclusively determined. The proposed changes to the duty and rest requirements of pilots will likely mean airlines need more pilots to comply with the new rules. The officials said the industry will likely see a greater demand for these additional pilots over the next few years, but this will depend on each airline's staffing ratios for its aircraft and flight operations. However, the general response was that the rule changes would require more qualified pilots in order to maintain the same level of service.

End Notes

[1] Regional air carriers are airlines that generally operate aircraft seating fewer than 90 passengers, and mainline carriers operate aircraft seating 90 or more passengers.

[2] Regional Airline Association, 2010 Annual Report, September 2010.

[3] NTSB Aircraft Accident Report, 2010, Loss of Control on Approach, Colgan Air, Inc., Operating as Continental Connection Flight 3407, Bombardier DHC-8-400, N200WQ, Clarence Center, New York, February 12, 2009. NTSB/AAR-10/01. Washington, D.C.

[4] The U.S. Department of Transportation Office of Inspector General is conducting an audit of FAA's oversight of air carrier pilot training programs at the request of Congress.

[5] Pub. L. No. 111-216, 124 Stat. 2348 (2010).

[6] Federal aviation regulations are found under title 14 of the United States Code of Federal Regulations. Part 135 of Title 14 prescribes rules governing the commuter or on-demand operations to hold an air carrier certificate. Nonscheduled-service airlines are generally issued a Part 135 certificate by FAA and operate aircraft other than turbojet-powered airplanes having no more than nine passenger seats and a payload capacity of 7,500 pounds or less. Part 121 prescribes rules governing the domestic, flag, and supplemental operations to hold an air carrier certificate. Scheduled-service airlines are generally issued a Part 121 certificate by FAA and operate turbojet-powered airplanes or airplanes with more than nine passenger seats or airplanes having a payload capacity of more than 7,500 pounds.

[7] Part 61 prescribes the minimum training, knowledge, and experience requirements for acquiring a pilot certificate. Part 141 prescribes the requirements for issuing pilot school certificates and the general operating rules applicable to a holder of the certificate. Part 142 prescribes the requirements governing the certification and operation of aviation training centers.

[8] 14 C.F.R. Part 61.

[9] The student pilot certificate is necessary to begin pilot training to work toward a recreational, sport, or private pilot certificate under the direct supervision of a flight instructor. The recreational pilot certificate was introduced in 1989 and limits pilots to stay within 50 nautical miles of their departure airport unless other conditions are met. The sport pilot certificate was introduced in 2004 and allows pilots to fly smaller, light, less complex, one or two-seat airplanes without the limitations of recreational pilots. The private pilot certificate allows the pilot to fly solo and carry passengers in any aircraft for which they are qualified, but not to fly for compensation. The commercial pilot certificate allows a pilot to be compensated for flying and to be hired for a variety of aviation jobs (e.g., air ambulance, law enforcement, agricultural spraying, corporate and charter flights, and banner towing), including first officer positions with airlines. The airline transport pilot certificate is the highest level of pilot certification, requires the highest amount of cumulative flight time, and is necessary to qualify as a captain for an airline.

[10] The holder of a valid flight instructor certificate may provide pilot training and instruction for the pilot certification in any aircraft for which they are qualified. (14 C.F.R. § 181).

[11] The DOT Office of Inspector General has initiated reviews related to FAA's oversight of airline pilot training, pilot fatigue, and professionalism issues, and whether there is a direct

relationship among commercial aviation accidents, pilot experience, and pilot compensation.

[12] A rating defines the conditions or specific aircraft in which a pilot certificate may be used. In addition, endorsements may be issued by FAA to further define conditions or specific aircraft not covered by ratings.

[13] Part 61 requires that pilots must be able to read, write, speak, and understand the English language, or a pilot certificate cannot be issued. Each pilot applicant must demonstrate an ability to mentally grasp critical information that often must be read and understood while conducting a variety of aviation operations.

[14] A pilot with a commercial pilot certificate can be employed to tow banners and fly sightseeing operations or can be employed as a first officer for a regional airline.

[15] In commercial aviation, the pilot in command (captain) of an aircraft is the person aboard the aircraft who is ultimately responsible for its operation and safety during all phases of flight, as well as when it is operating or moving on the ground, in accordance with FAA's rules of the air. The second-in-command (first officer) is the second pilot of an aircraft, and has the authority to assume command of the aircraft in the event of incapacitation of the captain. However, control of the aircraft is normally shared equally between the captain and first officer during flight.

[16] To be eligible for a student pilot certificate, an applicant must be at least 16 years of age and hold at least a third-class medical certificate. Student pilots are generally not allowed to operate an aircraft in solo flight unless certain FAA requirements are met.

[17] FAA appoints individuals as designated pilot examiners to conduct various pilot certification-related activities on behalf of FAA.

[18] Required inspections are identified by FAA's national office as the mandatory core inspections necessary for FAA to fulfill its statutory and regulatory oversight responsibilities. Accomplishment of these inspections is essential to provide reasonable assurance of continued compliance with regulations, standards, and safe operating practices. Planned inspections are identified at the regional and district levels and are comprehensive targeted inspections. They make up the depth and substance of each office's annual work program and FAA district offices tailor them to the changing local aviation environment.

[19] PTRS was implemented in 1990 and is one of FAA's data management and analysis systems used, among other things, to record inspections of flight instructors, pilot schools, pilot examiners, and other job functions.

[20] EIS is the primary internal tool for FAA in initiating enforcement actions and for tracking of enforcement and compliance cases and their outcomes.

[21] ICAO members, including the United States, are not legally bound to act in accordance with ICAO standards and recommended practices. However, if members decide to deviate from some of the ICAO standards and recommended practices, they are obligated to notify ICAO of differences. ICAO is a sovereign body, consisting of 190 members. Each member is entitled to one vote, and decisions are determined by majority vote.

[22] In August 2011, FAA issued a rule that, among other things, will allow vocational pilot schools to use internet-based training programs without requiring the schools to have a physical ground training facility. The rule will become effective October 31, 2011. 76 Fed. Reg. 54095 (August 31, 2011).

[23] 14 C.F.R. § 141.55.

[24] Training under Part 61 regulations for the private and commercial pilot certificates can accomplish the ground school training requirements through instruction from a flight instructor or through an approved home-study course.

[25] For the airline transport pilot certificate, the same number of total hours is required whether pilot training is being conducted under Part 61 or Part 141 regulations.

[26] Additionally, the Transportation Security Administration requires pilot schools to verify a student's proof of U.S. citizenship.

[27] Institutional accreditors review and accredit entire educational institutions. Program accreditors review and accredit specific programs or subject area offerings within an educational institution, such as the Aviation Accreditation Board International (AABI). AABI is an independent body that advances aviation education through accreditation and is the only program accreditor for collegiate, non-engineering aviation education for both 2-year and 4-year programs.

[28] The Flight School Association of North America is a membership-based association that represents pilot schools and firms that provide products or services to the pilot training or aviation industry.

[29] Some vocational and collegiate aviation schools with a Part 141 certificate are granted examining authority by FAA, as a privilege of their Part 141 certificate, which allows them to recommend their graduates for some of the initial pilot certificates and ratings (except flight instructor certificates, airline transport pilot certificates and ratings, and jet type ratings) without the pilot candidates having to take the FAA knowledge or practical tests or both.

[30] NASA, "FAA Pilot Knowledge Tests: Learning or Rote Memorization?" Ames Research Center, Moffett Field, California, January 2004.

[31] FAA, International Journal of Applied Aviation Studies, Pilot Source Study: An Analysis of Pilot Backgrounds and Subsequent Success in US Regional Airline Training Programs (Oklahoma City: August 2010). The study was conducted by a consortium of aviation university researchers in response to FAA's Advance Notice of Proposed Rulemaking for soliciting comments and recommendations on whether FAA should consider crediting specific academic study in lieu of flight hour requirements. 75 Fed. Reg. 6164 (Feb. 8, 2010). FAA was subsequently required by law to consider allowing specific academic training courses to count toward the 1,500 flight hour requirement. Pub. L. No. 111-216, § 217(d), 124 Stat. 2348, 2368 (2010).

[32] The Regional Airline Association, founded in 1975, represents regional airlines and provides technical, educational, and promotional support to the 32 members and their supply company partners.

[33] U.S. mainline airlines generally require pilot candidates to have a college degree, but not necessarily an aviation-related degree.

[34] In practice, regional airlines do typically require newly hired first officers to have a certain level of experience above the regulatory flight hour minimums required to obtain a commercial pilot certificate, typically hiring pilots that have spent several years working up through commercial non-airline jobs, i.e., on-demand and cargo operations. However, the extent that they do this may vary based on the cyclical dynamics of the supply of and demand for pilots.

[35] A flight engineer is the third crewmember on an aircraft that requires a three-person flight crew, but this position is seen less often as airlines replace older planes with newer aircraft that require only two-person crews.

[36] According to literature we reviewed, scenario-based training uses real-world situation that introduces situations and circumstances that pilots face in routine flight operations as learning experiences. This training method emphasizes the development of critical thinking and flight management skills, rather than solely on traditional maneuvers-based training

skills. Maneuvers-based training emphasizes the mastery of individual tasks or elements, such as the development of competency and motor skills to satisfactorily accomplish individual piloting maneuvers.

[37] The FAA Industry Training Standards program (referred to as FITS) is a partnership between FAA, industry, and academia to create scenario-based training materials to help pilot schools train pilots for practical application of knowledge and skills of technically-advanced aircraft. The FITS commercial pilot syllabus, developed in 2007, is the accepted training method to use in developing a specific FITS curriculum.

[38] However, recent concerns have been raised by industry stakeholders that federal regulations and airline operating procedures require airline pilots to have a greater reliance on the computerized flight management and automation of today's aircraft. As a result, the implications of decreased opportunities for pilots to manually fly an aircraft could potentially result in an increase in accidents and incidents for pilots who are suddenly confronted with a loss of computerized controls or emergency situations but unprepared to respond immediately or make errors in doing so.

[39] Advanced jet training is designed to give instruction in airline flight crew operations in a multiengine aircraft, emphasizing the transition of the professionally qualified pilot to a highly skilled member of an airline flight management team. Course topics include crew resource management, flight crew training techniques, high speed and high altitude programming of automatic flight control systems, transport aircraft flight techniques, turbojet operations in all flight regimes and in difficult operating conditions, and use of advanced avionics.

[40] Aircraft upset is a dangerous condition that occurs when an airplane encounters an unexpected event that threatens loss of control of the aircraft and may result in an accident. Upset recovery training aims at improving a pilot's ability to maneuver and recover an actual airplane from a serious upset to a normal operating condition.

[41] Crew (or cockpit) resource management training is designed to improve pilots' team management, situational awareness, communications, task allocation, and decision-making skills in an airline crew environment.

[42] Pub. L. No. 111-216, §217 (c) (1), 124 Stat. 2348, 2368 (2010). 4314 C.F.R. § 61.129(b).

[44] Pub. L. No. 111-216, §217(c)(1),124 Stat. 2348,2368 (2010). 45Pub. L. No. 111-216, §217(f), 124 Stat. 2348, 2368 (2010).

[46] FAA, The First Officer Qualifications Aviation Rulemaking Committee Report: Recommendations Regarding Rulemaking on Flight Experience, Training, and Academic Requirements Prior to Operating as a First Officer in Part 121 Air Carrier Operations, September 10, 2010. The committee was composed of representatives of AABI, ALPA, Air Transport Association, AOPA, Coalition of Airline Pilots Associations, National Air Disaster Alliance-Foundation, National Business Aviation Association, Pilot Career Initiative, Regional Air Cargo Carriers Association, and RAA.

[47] CAPA is a trade association, established in 1997, comprised of over 28,000 professional pilots to address safety, security, legislative and regulatory issues affecting the individual member unions. The National Air Disaster Alliance-Foundation was founded by air crash survivors and victims' family members in 1995, to raise the standard of safety, security, and survivability for aviation passengers, and to support victims' families.

[48] A bridge agreement is a formal signed document between an aviation program and a regional airline that generally reduces the minimum flight time experience required by the airline for students to be considered for hire. Typical conditions of the agreement stipulate a specific

grade point average, minimum number of flight hours, and other desirable academic qualifications for the students.

[49] Renewal of a flight instructor certificate can be completed in three ways: (1) passing a practical test for renewal of the flight instructor certificate, (2) meeting required FAA minimum of flight activity conducted as an instructor, or (3) completing a flight instructor refresher course consisting of ground training or flight training or both.

[50] Runway incursions, which are precursors to aviation accidents, are instances where collisions are narrowly avoided on an airport's runways or taxiways. For more information on runway incursions, see GAO, Aviation Safety: FAA Has Increased Efforts to Address Runway Incursions, GAO-08-1169T (Washington, D.C.: September 2008).

[51] UAA is the representative voice of college aviation education to the aviation industry, government agencies, and the general public.

[52] 76 Fed. Reg. 29336 (May 20, 2011). The original notice of proposed rulemaking was published in January 2009. 74 Fed. Reg. 1280 (Jan. 12, 2009).

[53] Pub. L. No. 111-216, § 208, 124 Stat. 2348, 2360 (2010).

[54] Flight Safety International is a global professional aviation training company that operates the world's largest fleet of advanced full flight simulators and supplier of flight simulators, visual systems, and displays.

[55] EASA is the European aviation regulatory authority and FAA counterpart. EASA became operational in September 2003 on the basis of European Parliament and Council regulation 1592/2002. The agency is an independent European Union body under European law accountable to the Member States and the European Union institutions. In February 2008, Regulation 216/2008 was passed repealing Regulation 1592/2002 and included the scope of oversight for flight crew licensing.

[56] Ab initio is a Latin term meaning "from the beginning." Ab initio pilot training programs trains students with little to no pilot experience to become commercial airline pilots through a continuous training program.

[57] Approved flight training organizations are required to meet certain qualifying standards of personnel licensing.

[58] Competency-based training and assessment is defined as performance orientation with an emphasis on standards of performance and their measurement, and the development of training to the specified performance standards.

[59] In November 2006, ICAO enabled the implementation for the multi-crew pilot license by amending ICAO personnel licensing standards to include a new pilot certificate and adopting new standards for this ab initio airline pilot training method.

[60] A ramp check for a pilot school with a 141 certificate typically involves an inspection of the paperwork and exterior of the aircraft used by the school for training to ensure each meets all regulatory requirements. FAA inspectors conduct various types of ramp checks to ensure that aviation activities are being conducted safely and in compliance with the Federal Aviation Regulations.

[61] FAA uses a broad pool of designated personnel nationwide to act as representatives of the agency to conduct many safety certification activities, such as administering practical tests to pilots and many other activities.

[62] Part 61 provides the regulatory basis for the conduct of practical tests for pilot certification. Section 183.23 of title 14 (subpart C) provides the regulatory basis for the designation of pilot examiners. FAA Order 8900.2 contains procedures for the selection, appointment, oversight, training, renewal, termination, and appeal of designees.

[63] High activity pilot examiners are defined as conducting 50 or more practical tests per quarter.

[64] These circumstances may include a designee: (1) who does not comply with the applicable certification policies; (2) whose certification passing rate is excessively high; (3) who exceeds the allowable amount of certification activity on a given day; (4) who conducts any certification activity that requires prior permission from the managing FAA office without obtaining that permission in advance; (5) whose certification file error rate is excessively high; (6) who is the subject of a valid public complaint; or (7) who has been involved in an accident, incident, or violation of federal aviation regulations.

[65] Low-activity pilot examiners were those with seven or less activities between the third quarter of 2010 and the second quarter of 2011.

[66] SPAS relies on over 25 databases within FAA and other government agencies, including PTRS, EIS, and the Vital Information Subsystem, which maintains a current list and the status of pilot school certificates.

End Notes for Appendix II

[1] The Aircraft Owners and Pilots Association is the largest not-for-profit general aviation association in the world and serves more than 415,000 members.

[2] Pub. L. No. 110-135, §2, 121 Stat. 1450 (2010).

[3] In August 2009, FAA issued a final rule that increased the duration of validity for student pilot certificates for pilots under the age of 40 from 36 to 60 months. 74 Fed. Reg. 42500 (Aug. 21, 2009). This resulted in the increase in active student pilots to 119,119 from 72,280 at the end of 2009.

[4] AOPA, The Flight Training Experience: A Survey of Students, Pilots, and Instructors, October 2010.

[5] The Federal Family Education Loan program allowed private financial institutions to provide students with loans, but the federal government assumed the risk of default, and paid the financial fees while the student attended college. As of July 2010, federal education loans are now made directly by the U.S. Department of Education under the Federal Direct Loan Program.

[6] Sallie Mae is the largest servicer of federally guaranteed student loans and the largest originator and servicer of private student loans.

[7] GAO, Higher Education: Factors Lenders Consider in Making Lending Decisions for Private Education Loans, GAO-10-86R (Washington, D.C.: November 2009).

[8] Pub. L. No. 111-216, §217(a), 124 Stat. 2348, 2367.

[9] 75 Fed. Reg. 55852 (Sept. 14, 2010). FAA's proposed rule (1) sets a 9-hour minimum for rest prior to flying-related duty (a 1-hour increase over the current rules), (2) establishes a new method for calculating a pilot's weekly, monthly, and yearly flight time limits so that pilots can receive at least 8 hours of sleep during that rest period, and (3) sets a daily flight duty period to 13-hours (currently it is a 16-hour duty period between rest periods).

[10] Pub. L. No. 111-216, §212, 124 Stat. 2348, 2362.

In: Pilot Training and Safety
Editor: David M. Fontenelle

ISBN: 978-1-62417-400-1

Chapter 2

STATEMENT OF CALVIN L. SCOVEL III, INSPECTOR GENERAL, U.S. DEPARTMENT OF TRANSPORTATION. HEARING ON "COMMERCIAL AIRLINE SAFETY OVERSIGHT"*

Madam Chairman and Members of the Subcommittee:

Thank you for inviting me here today to testify on the Federal Aviation Administration's (FAA) progress in implementing advanced standards for pilot training and new safety measures for air carriers. As you know, following the 2009 Colgan Air crash, Congress and FAA took swift action to implement measures for improving pilot training and qualifications programs, reducing flight crew fatigue, and ensuring operators meet safety standards. These efforts culminated with the August 2010 passage of the Airline Safety and FAA Extension Act,[1] which contains new requirements to enhance safety in these areas. Effective implementation of these requirements should go a long way in improving safety in commercial airline travel.

Prior to passing the Act, this Subcommittee, as well as the House Committee on Transportation and Infrastructure, requested that we review FAA and industry efforts to enforce new Federal regulations for flight crew rest requirements and address fatigue issues and airline pilot training program

* This is an edited, reformatted and augmented version of Statement Presented March 20, 2012 before the Senate Committee on Commerce, Science, and Transportation, Subcommittee on Aviation Operations, Safety & Security.

weaknesses. Since we began our work in 2009, we have issued three reports.[2] Our testimony today is based on those reports as well as our ongoing work regarding implementation of the Act. Today, I will focus on (1) FAA's progress in responding to provisions of the Act, (2) the challenges FAA faces in implementing certain provisions, and (3) concerns related to achieving the full measure of safety enhancements intended by the Act.

In Summary

FAA has met or is on schedule to meet many of the Act's requirements, such as improving pilot rest requirements and establishing better processes for managing safety risks.

However, FAA has not met timelines for raising pilot training standards, implementing mentoring programs, providing enhanced leadership skills to captains, and increasing minimum pilot qualifications. FAA also faces challenges in establishing a pilot records database—an important component for enhancing the air carrier screening process for pilot applicants. In addition to overcoming these challenges, FAA needs to provide additional guidance and assistance to industry— especially smaller carriers—in developing and managing new safety programs.

Background

The 2010 Act included 16 provisions to improve airline safety and pilot training with milestones spread over a 3-year period. The Act called for advanced standards for pilots, including required rulemaking activities for training programs, crewmember screening and qualifications, and new fatigue regulations to improve passenger safety.

These rulemaking activities are complex, and some have encountered significant air carrier opposition. In addition to notice and comment periods required by law, FAA must conduct detailed analyses of each rule's likely effects and coordinate with stakeholders.

The Act also included several important initiatives that FAA did not complete during its Call to Action on Airline Safety,[3] such as developing mentoring and professional programs for pilots and following up with air carriers on efforts to adopt voluntary safety programs. In addition, the Act requires FAA to establish a pilot records database that air carriers must access to review qualifications and past performance data before hiring pilots. (See exhibit for further detail and current status of FAA's efforts in each section of the Act.)

FAA Met Act Requirements to Address Pilot Fatigue and Advanced Some Safety Initiatives at Air Carriers

FAA developed a concerted strategy to meet the Act's stringent timelines and implement new safety programs, including issuing a final rule on crew rest and fatigue, increasing air carrier use of voluntary safety programs, and advancing Safety Management Systems (SMS).

FAA Overhauled Flight and Duty Time Regulations

In January 2012, FAA updated its flight and duty time regulations for Part 121[4] air carrier pilots to better ensure pilots are rested when they fly. This is a significant achievement for the Agency given that these updates were the first modifications to the regulations since 1985 and that the proposed rule received over 8,000 comments from the aviation industry, mostly opposing the planned requirements.

Unlike the old rules—which included different rest requirements for domestic, international, and unscheduled flights—the new regulations establish one set of rules that are based on scientific factors, such as the time of day pilots begin their first flight, the number of scheduled flight segments, and the number of time zones crossed.

Pilots are also now required to affirmatively state that they are fit to fly and are prohibited from flying during a scheduled duty period when they report fatigue.

Other key changes in the new flight and duty time regulations include a 10-hour minimum rest period prior to duty, a 2-hour increase over the previous rule, and 30 consecutive hours free from duty per week—an increase of 25 percent over the previous regulation requirements.

FAA Promoted Air Carriers' Use of Voluntary Safety Programs

In March 2011, FAA completed a congressionally required review of Part 121 air carriers' use of voluntary safety programs[5] and later devised a plan to help smaller air carriers implement these programs. Data gathered through voluntary safety programs can be used to identify the trends and patterns that represent risks. The Act targets air carrier participation in three such programs that FAA oversees:

- *Aviation Safety Action Program (ASAP)*, which encourages air carrier and repair station employees to voluntarily report safety information that may be critical to identifying potential precursors to accidents without fear of enforcement or disciplinary action.
- *Flight Operational Quality Assurance (FOQA)*, which collects and analyzes digital flight data generated during scheduled flights to provide greater insight into performance and operations.
- *Advanced Qualification Program (AQP)*, which provides a voluntary alternative to traditional training standards by incorporating data-driven quality control processes to refine pilot training based on the individual's proficiency and identified training needs.

The review found that, as of March 2011, 68 percent of Part 121 air carriers participate in at least one voluntary safety program and just under half of those carriers use more than one. Our ongoing analysis of current FAA data as of January 2012 shows a continued rise in voluntary safety program use—70 percent of Part 121 air carriers have at least one program, up from 59 percent 2 years ago. Further, for the same time period, 47 percent of Part 121 air carriers have multiple programs, compared to 36 percent 2 years ago.

As shown in figure 1 below, the highest concentration of new growth for these air carriers has been with ASAP and FOQA. We also determined that the majority of carriers that transitioned to AQP training were regional carriers. Air carriers that use AQP benefit from improved data collection and analysis techniques and enhanced flight crew coordination.

FAA Is Ahead of Schedule for Implementing SMS

Increasing use of voluntary safety programs is important for Part 121 air carriers of all sizes, as the data generated by these programs is a large driver of

SMS, a systemic and comprehensive process for managing safety risks. Specifically, SMS provides operators with business processes and management tools to examine data from everyday operations, isolate trends that may be precursors to incidents and accidents, and develop and carry out appropriate risk mitigation strategies. FAA has nearly completed its efforts to issue a final rule on SMS for air carriers. The Agency released a proposed rule in October 2010 and, according to the Act, has until August 2012 to issue a final rule.

FAA has also taken steps to assist air carriers in developing SMS before the completion of the final rule. FAA developed an SMS pilot program in 2007 to develop implementation strategies and oversight responsibilities. SMS pilot projects allow FAA and air carrier input in developing guidance and provide carriers an opportunity to share best practices and lessons learned. Currently, 83 percent of all Part 121 air carriers (73 of 88) are participating in the pilot program.

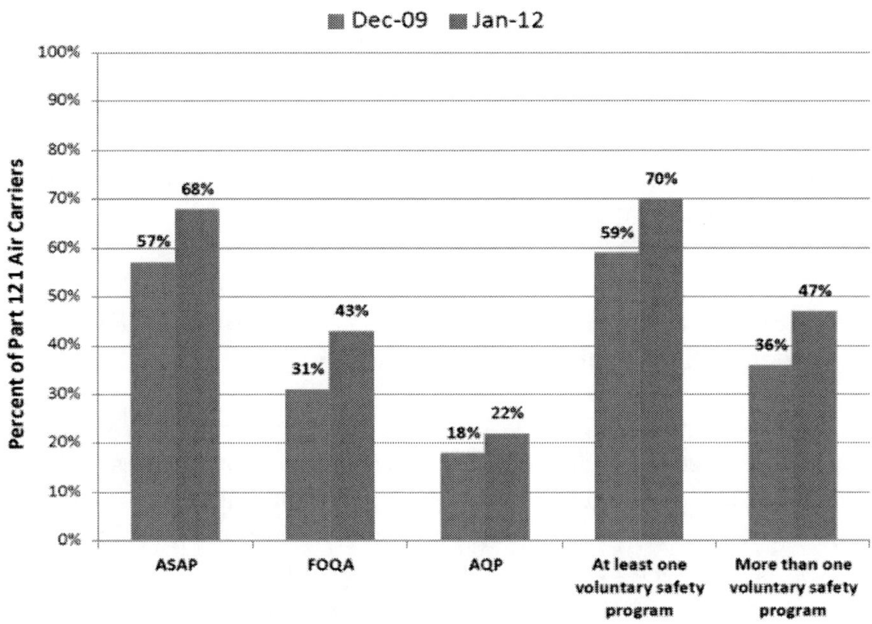

Source: OIG analysis of FAA-reported data.

Figure 1. Increase in Voluntary Safety Program Participation.

The new system, when fully implemented across all carriers, has the potential to significantly advance safety. However, there is industry concern that the SMS rule will not be scalable for air carriers of varying size and operational complexity, posing a larger burden on smaller air carriers for their implementation. Currently, 14 of the 15 carriers that are not yet participating in FAA's SMS pilot program are smaller carriers (with less than 20 aircraft). Additionally, air carriers are concerned about public disclosure of SMS-collected data. Most of these concerns focus on whether the data can be used in legal proceedings. The current proposed rule does not address these concerns.

FAA Faces Challenges in Meeting Act Provisions on Pilot Training and Safety Issues

FAA efforts to issue Act-mandated rules to improve training standards, establish mentoring and leadership programs, and enhance screening and qualifications for pilots are delayed. While FAA is responsible for raising pilot qualifications, the successful implementation of such rules depends on FAA's ability to address air carrier concerns and work through the regulatory process in a timely manner, which has been a significant challenge for FAA. The Agency also faces several challenges in establishing a new centralized electronic pilot records database to provide air carriers with better background information on pilots they intend to hire.

Industry Concerns Have Delayed FAA's Rulemaking Efforts to Enhance Pilot Training Standards

FAA is almost 6 months overdue on issuing a final rule revising pilot training requirements—the delay is due in part to significant industry opposition. This is an important safety initiative that will require pilot training programs to incorporate flight simulators and enhance pilots' ability to work together during emergencies. In January 2009, FAA issued the Notice of Proposed Rulemaking (NPRM). However, FAA received extensive industry comments, primarily opposing that the rule imposes overly prescriptive training hours rather than basing pilot training on skills most needed to safely perform flight operations. As a result, FAA issued a second proposed rule in May 2011. The revised proposal requires more thorough ground and flight training for pilots on how to recognize and recover from stalls, as well as remedial training for pilots who perform poorly in training.

With advancements in pilot training on the horizon, it is important that FAA enhance its oversight practices. For example, under the new rule, carriers will be required to provide remedial training for pilots with performance deficiencies. However, it will be difficult for FAA to gauge the effectiveness of this training unless it corrects weaknesses we identified in our December 2011 report on pilot training. We found that FAA was not tracking poorly performing pilots due to inadequate guidance for its inspectors on how to gather data on pilot performance. Specifically, FAA guidance requires inspectors to compare pilot proficiency checks that they have performed against those conducted by the carriers' check airmen.[6] However, we questioned the viability of this requirement since nearly all pilot proficiency checks are conducted by check airmen, not FAA inspectors. As a result, FAA inspectors may not have sufficient data to make a meaningful comparison.

FAA's Rule to Require Pilot Mentoring, Leadership, and Professional Development Committees at Air Carriers Is Overdue

FAA is also almost 8 months overdue in meeting a mandated timeline to issue a proposed rule requiring that air carriers establish pilot mentoring, leadership, and professional development committees to improve pilot performance. This is due, in part, to a lengthy delay in developing an appropriate balance between the costs and benefits of these programs. FAA intends to issue a proposed rule that it believes would generate benefits by reinforcing safe flying practices.

Pilot performance issues are longstanding safety concerns—pilot performance was cited in 7 of the 10 major accidents that occurred over the last decade, indicating that the quality of training, professionalism, and mentoring is important to safety. In February 2011,[7] we also reported that poor pilot performance—such as poor decision-making, inadequate aircraft control, improper flying techniques, and a disregard for operating procedures—is a high-ranking causal factor in airline accidents,[8] a finding consistent with the National Transportation Safety Board (NTSB) review of the Colgan accident.

After the Colgan accident, FAA focused on advancing pilot mentoring as part of its 2009 Call to Action on Airline Safety and Pilot Training but never released a detailed plan to implement programs at air carriers. In December 2011, we reported that regional air carriers were not pursuing mentoring opportunities for their pilots. Officials at these carriers expressed concerns that a mentoring program would have to be scaled to their business model and that pilot turnover at these carriers could outweigh the benefits of establishing these often costly programs. Seven of nine large and small carriers we recently visited as part of our ongoing audit did not have formal mentoring programs, and none had professional development programs to instill and reinforce high performance standards for their pilots.

FAA Is Behind Schedule and Will Likely Miss the Deadline to Issue a Rule Increasing Minimum Qualifications for Commercial Airline Pilots

The Act requires FAA to issue a final rule to substantially raise airline pilot qualifications by August 2012. However, FAA did not issue the proposed rule until February 2012 and expects to issue the final rule by August 2013—a year after the mandate. Given the significant increase in pilot flight hours that the Act mandates for the final rule, FAA has encountered industry opposition.

FAA's rule would require first officers to hold an Airline Transport Pilot (ATP) certificate,[9] requiring 1,500 hours of pilot flight time—up from the current mark of 250 hours with a commercial pilot's license. The proposed rule would also require first officers to have an aircraft type rating, which involves additional training and testing specific to the airplanes they fly. Air carrier representatives are opposed to the increased flight hour requirement because they feel a pilot's quality and type of flying experience should be weighted more heavily than the number of flight hours. They state that the supply of qualified and available pilots will decrease because it will be difficult for entry-level pilots to attain this amount of hours before being qualified to fly at a commercial air carrier.

Further, while FAA's pilot qualification proposal satisfies most of the Act's requirements in this area, it may fall short in ensuring sufficient pre-employment screening. For example, the Act states that applicant pilot screening must include an assessment of skills, aptitudes, airmanship, and suitability specific to each air carrier's operations. However, it is unclear whether FAA intended for the enhanced ATP requirements in the proposed rule to also satisfy the pre-employment screening measures contained in the Act. If so, air carriers may not make appropriate changes to their pre-employment screening procedures specific to their operations.

Finally, FAA has not acted to ensure carriers are ready to transition to these new pilot qualification requirements. For example, at two regional air carriers we visited as part of our ongoing review, more than 75 percent of current first officers did not have an ATP. Yet, neither carrier had developed a plan to ensure these pilots would be able to meet the enhanced requirements by the deadline, nor had the local FAA inspectors followed up with these carriers to assess their ability to comply with enhanced requirements. Additionally, FAA has not taken steps to determine the potential impact the new ATP requirement would have on current pilots, information that will be important for safety oversight.

FAA Lacks a Clear Strategy for Transitioning to a New Centralized Electronic Pilot Records Database

FAA met the Act's initial milestone in developing a centralized electronic pilot records database that will include records previously maintained by air carriers.

The Act did not prescribe any additional milestones for the database's implementation, but the Agency has recognized that a rulemaking will be necessary to fully develop the intricacies of this electronic system and is in the preliminary stages of writing this proposal.

However, to create a robust, complete, and secure data repository that carriers can use when hiring pilots, FAA must overcome three key challenges:

- First, FAA must address what level of detail should be captured from air carrier pilot training records, such as whether recurrent flight training will be included. The Act stipulates that comments and evaluations made by the pilot examiner10 be included in the database; however, industry is highly protective of these data and opposes including them in the database. FAA must also address how to include historical air carrier pilot training records into its new system. Gathering the historical records while keeping them standardized across sources will be difficult because information in the records varies based on differences in air carrier training programs and the record retention period varies from 5 years to indefinitely, depending on the carrier.

- Second, FAA does not expect to issue a final rule and launch the database for at least another 2 years, so FAA will have to determine how to transition from current recordkeeping practices mandated by the Pilot Records Improvement Act (PRIA)11 to the new database without disrupting the flow of information. Therefore, until air carrier records are fully integrated into the new database, carriers may need to continue requesting data from both FAA and previous employers.
- Finally, a pilot records advisory committee identified multiple challenges for FAA in accessing records from the National Driver Register (NDR)12 and incorporating them into the database. For example, FAA must decide how to ensure data reliability of pilot records and resolve conflicting data retention policies for the database versus NDR.

In addition to these challenges, we reported in December 2011 that FAA lacks a centralized process to receive and respond to carriers' requests for pilot records.

This raises questions about whether air carriers are getting all the relevant information FAA has on pilots before they are hired. While this problem may be mitigated once the new database is launched, it remains a concern in the interim, especially since FAA's 2009 Call to Action on Airline Safety and Pilot Training called on carriers to obtain more comprehensive records on pilots prior to hiring. As a result, FAA experienced an influx of record requests from carriers and an increased workload.

Sustained Commitment and Oversight Are Needed to Achieve the Full Measure of Safety Enhancements Intended by the Act

Sustained FAA management commitment and oversight are needed to ensure that provisions of the Act are effectively implemented and have the desired impact of improving safety. Specifically, FAA needs to (1) effectively communicate with local FAA offices and industry on the status of new rules and guidance, (2) provide additional direction and support for developing new safety programs at smaller air carriers, and (3) address pilot commuting issues.

FAA Has Not Provided the Level of Education, Outreach, and Guidance Needed for Industry to Implement New Safety Programs

A lack of key stakeholder involvement and poor communication between FAA and industry is impeding progress on several Act initiatives. FAA created six aviation rulemaking committees (ARC) to develop recommendations on multiple initiatives, such as identifying and promulgating best practices in pilot training and developing the pilot record database. However, FAA did not inform its field offices or airlines of many of the ARCs' results, such as status of rulemakings, or engage in effective outreach efforts for new safety programs other than SMS.

For example, none of the nine field offices we visited during our ongoing review had received information from FAA on the Agency's progress in developing mentoring, professional development, and leadership programs for air carrier pilots.

If FAA had provided early outreach to field offices on the status of rulemaking and best practices, air carriers could be better positioned to implement new pilot safety programs when the rules are finalized.

Further, FAA did not follow up to ensure proper implementation of guidance it has issued to air carriers. For example, while FAA issued guidance for retaining and submitting pilot training records for the new electronic, centralized pilot records database, it did not follow up to see that air carriers were following the new requirements.

Four of the six carriers we visited during our ongoing review had not clarified their policies to reflect this change. As a result, important details concerning pilot training and proficiency may be lost and not available for air carriers to use in future hiring decisions.

Additionally, some air carriers that had moved forward with new programs encountered obstacles in obtaining FAA approval. For example, one regional air carrier attempted to proactively develop a program for first officers to obtain advanced certification as prescribed by the Act.

While the local FAA office initially approved the program, FAA rescinded the approval 1 day before it was set to launch because national-level guidance had not been issued.

Although most actions taken by air carriers thus far are voluntary, a lack of clear and timely communication by FAA provides little impetus for air carriers to move forward with new initiatives.

FAA Does Not Have a Focused Plan to Help Smaller Air Carriers Establish New Safety Programs

Despite overall gains, implementation of voluntary safety programs has mostly occurred at larger air carriers. Yet, the Act instructed FAA to develop a plan to help all Part 121 carriers establish such programs, with particular emphasis on ASAP and FOQA. FAA devised an implementation plan for ASAP and FOQA at smaller air carriers, but a lack of funding has prevented FAA from enacting the FOQA plan. As a result, smaller carriers have been unable to purchase and install the equipment needed to run this program. With a focused plan and dedicated funding, FAA can help smaller air carriers establish voluntary safety programs and realize the benefits of increased safety reporting and trend analyses.

As shown in table 1, while all large carriers with more than 50 aircraft in their fleet have an incident reporting system (ASAP), the system has been adopted by only 41 percent of small carriers with 15 or fewer aircraft. Similarly, just over 10 percent of these small operators have FOQA, compared to more than 90 percent of large operators.

Table 1. Air Carrier Voluntary Safety Program Participation

Program	Number of Carriers Participating	Large Carriers (more than 50 aircraft)	Medium Carriers (16-50 aircraft)	Small Carriers (15 or fewer aircraft)
Aviation Safety Action	60 of 88	24 of 24	19 of 23	17 of 41
Program	(68%)	(100%)	(83%)	(41%)
Flight Operational	38 of 88	22 of 24	11 of 23	5 of 41
Quality Assurance	(43%)	(92%)	(48%)	(12%)
Advanced Qualification	19 of 88	13 of 24	3 of 23	3 of 41
Program	(22%)	(54%)	(13%)	(7%)

Source: OIG analysis of FAA-reported data as of January 2012.

Despite the disparities between the large and small air carriers, FAA has not fully implemented its plan to assist smaller air carriers with the resources needed—such as best practices and guidance—to establish new safety programs. Smaller air carriers have fewer resources than their mainline counterparts to handle the operation and management of new safety programs. As a result, they will have to prioritize development of these programs based on feasibility and importance and will face difficulties in implementing new programs simultaneously—especially without guidance or program assistance from FAA.

FAA's Fatigue Rule Does Not Address Pilot Commuting

FAA's changes to the flight and duty time regulations represented a significant safety achievement; however, the regulations do not require air carriers to identify pilots who commute. These are significant factors that may contribute to fatigue given that many pilots in the industry reside hundreds or even thousands of miles from their duty locations. While FAA considered mandating that pilots arrive in time to receive a preflight rest period in the proposed rule, it stated that the requirement would be difficult to enforce and would not guarantee responsible commuting.

Pilot commuting and related issues were concerns that came to light after the Colgan accident. The NTSB investigation into the crash revealed that both pilots had commuted hundreds of miles before the flight. NTSB also found that Colgan did not proactively address the pilot fatigue hazards associated with basing its operations at an airport where pilots typically have to commute long distances in order to begin their work shifts. In its investigative report, the NTSB stated that "operators have a responsibility to identify risks associated with commuting, implementing strategies to mitigate these risks, and ensure that their commuting pilots are fit for duty."

NTSB issued a recommendation to FAA to address fatigue risks associated with commuting, including identifying pilots who commute. The National Academy of Sciences similarly noted in a July 2011 report that there are not enough data to determine the role commuting plays in contributing to fatigue or whether it should be regulated.[13] This underscores how collecting and analyzing these data could help FAA make well-informed decisions on commuting. In our September 2011 report on pilot fatigue, we recommended that FAA collect and analyze data regarding pilot commuting for all Part 121 flight crews and determine if additional changes are needed or if airlines need to take further mitigating actions in their fatigue management systems. In its response, FAA stated that it will review available data on pilot commuting and determine if additional data could offer added safety benefits. FAA committed to completing these actions by October 1, 2012.

Conclusion

FAA plays an integral role in maintaining the excellent safety record of the U.S. National Airspace System. FAA acted swiftly to address safety concerns highlighted by the Colgan crash and has since made commendable progress in meeting new Act requirements. FAA still faces several challenges, however, in updating pilot training and leadership programs, developing screening and qualifications standards, and ensuring carriers have the data they need to make sound hiring decisions. To effectively implement these initiatives in a timely manner, FAA must balance industry concerns with a sustained commitment to oversight. We are encouraged by FAA's progress to date and will continue to monitor its efforts to meet remaining Act requirements.

This concludes my statement. I would be happy to address any questions from the Chairman or Members of the Subcommittee at this time.

Exhibit. Status of key airline safety act requirements

Section	Initiative	Milestone	Deadline	Milestone Status
202	NTSB Recommendations Report	Report	Annual	Met, On-Target
203	FAA Pilot Records Database	Database Development	10/30/2010	Met
		Report	2/1/2012	Missed & Overdue
204	Air Carrier Safety & Pilot Training ARC	ARC Report	7/31/2011	Met
		ARC Report	7/31/2012	On-Target
205	FAA Inspector Staffing	Start OIG Review	5/1/2011	Met
206	Mentoring, Development, and Leadership	NPRM	8/1/2011	Missed & Overdue
		Final Rule	8/1/2013	To Be Determined
207	Crew Pairing and CRM	Study	8/1/2011	Completed Late – 8/26/2011
208	NTSB Training Recommendations	ARC Formation	11/29/2010	Met
		NPRM	8/1/2011	Met
		ARC report	11/30/2011	Completed Late- 3/7/2012
		Final Rule	8/1/2013	To Be Determined
209	FAA Rulemaking on Training	ARC Formation	9/30/2010	Completed Late – 11/16/2010

Section	Initiative	Milestone	Deadline	Milestone Status
		ARC Report	8/1/2011	Completed Late – 9/23/2011
		Final Rule	10/1/2011	Missed & Overdue
210	Code Share Ticket Disclosure	Amend 49 U.S.C. § 41712	N/A	Met
211	FAA Safety Inspections	Perform one per year	Annual	Met
212	Fatigue & Commuting	NPRM	2/1/2011	Met
		Final Rule	8/1/2011	Completed Late – 1/4/2012
		Risk Management Plans	11/1/2010	Met
		Start Study	9/30/2010	Met
		Preliminary Findings	1/30/2011	Met
		Report	6/30/2011	Met
213	Voluntary Safety Programs	Report	1/28/2011	Completed Late – 3/16/2011
214	ASAP & FOQA Implementation	Plans Issued	1/28/2011	Completed Late – 4/14/2011
		Plans Implemented	8/1/2011	FOQA Portion Overdue
215	Safety Management Systems	NPRM	11/1/2010	Met
		Final Rule	8/1/2012	On-Target
216	Screening & Qualifications	NPRM	1/28/2011	Completed Late - 2/29/2012
		Final Rule	8/1/2012	To Be Determined
		ATP	8/1/2013	To Be Determined
217	ATP Certification	Final Rule	8/1/2013	On-Target

Source: OIG analysis of FAA-reported data.

End Notes

[1] Airline Safety and Federal Aviation Administration Extension Act of 2010, Pub. L. No. 111-216, August 1, 2010.

[2] OIG Correspondence Number CC-2009-074, "Letter to Senators Rockefeller, Hutchinson and DeMint Regarding Commercial Aviation Accidents, Pilot Experience and Pilot Compensation," February 9, 2011. OIG Report Number AV-2011-176, "FAA and Industry Are Taking Action To Address Pilot Fatigue, but More Information on Pilot Commuting Is Needed," September 12, 2011. OIG Report Number AV-2012-027, "New Approaches Are Needed To Strengthen FAA Oversight of Air Carrier Training Programs and Pilot

Performance," December 20, 2011. OIG reports are available on our Web site: www.oig.dot.gov.

[3] FAA's Call to Action Plan, announced on June 24, 2009, consisted of 10 short- and mid-term initiatives to enhance pilot performance and training, increase air carrier participation in voluntary safety programs, and expand pilot records review. FAA also set goals to develop new safety oversight guidance to its inspectors, issue rulemakings on pilot fatigue and training, conduct regional safety forums to discuss industry best practices, and develop programs addressing pilot professionalism.

[4] 14 CFR Part 121, Operating Requirements: Domestic, Flag, and Supplemental Operations.

[5] Voluntary Safety Programs, Response to P.L. 111-216, Sec. 213, January 28, 2011.

[6] Pilots employed by air carriers who evaluate a pilot's proficiency during training and examinations.

[7] OIG Controlled Correspondence CC-2009-074, "Letter to Senators Rockefeller, Hutchison, and DeMint Regarding Commercial Aviation Accidents, Pilot Experience and Pilot Compensation," February 9, 2011.

[8] We analyzed the experience (i.e., total flight time and total make and model flight time in the accident aircraft) of pilots involved in 322 scheduled Part 121 passenger accidents that occurred from January 2000 through December 2009.

[9] Airline Transport Pilot (ATP) Certificate is the highest level of pilot certification. Pilots certified as ATP are authorized to act as pilot-in-command of an aircraft in commercial airline service. Additional eligibility requirements are contained in 14 CFR 61.153.

[10] An FAA inspector or air carrier pilot who is qualified, and permitted, to conduct flight checks or instruction in an airplane, in a flight simulator, or in a flight training device for a particular type airplane.

[11] Pub. L. No. 104-264, Section 502 (codified at 49 U.S.C. § 44703(h)-(j)).

[12] NDR is a central information system that allows states to electronically exchange information on licensed drivers through a computerized network.

[13] The National Academy of Sciences, The Effects of Commuting on Pilot Fatigue, ISBN 978-0-309-21696-8, July 6, 2011 (Response to P.L. 111-216, Sec. 212).

In: Pilot Training and Safety
Editor: David M. Fontenelle

ISBN: 978-1-62417-400-1

Chapter 3

TESTIMONY OF WILLIAM R. VOSS, PRESIDENT AND CEO, FLIGHT SAFETY FOUNDATION. HEARING ON "COMMERCIAL AIRLINE SAFETY OVERSIGHT"*

Chairman Cantwell, Ranking Member Thune, and distinguished members of the Subcommittee: My name is William Voss and I am the President and Chief Executive Officer of the Flight Safety Foundation.

The Flight Safety Foundation is an independent, nonprofit, international organization engaged in research, education, advocacy, and publishing. Its mission is to be the leading voice of safety for the global aerospace/aviation community.

We have members all around the world representing every facet of the aviation industry.

On behalf of the Foundation, I appreciate this opportunity to testify about the implementation of HR 5900, the Airline Safety and Federal Aviation Administration Extension Act of 2010.

The short answer is that the implementation of HR 5900 is progressing well.

The more complicated answer is that there is still work to be done and parts of this law that ought to be reconsidered.

---

* This is an edited, reformatted and augmented version of a Testimony Presented March 20, 2012 before the Senate Committee on Commerce, Science, and Transportation, Subcommittee on Aviation Operations, Safety & Security.

THE FATIGUE RULE

My background in aviation is diverse. I've been a pilot, an air traffic controller, a certified aviation mechanic, and a regulator and standard-setter, both at the FAA and the International Civil Aviation Organization (ICAO). I know firsthand how the issue of fatigue can affect every aspect of this industry. We all know that fatigue affects our performance, but normally our own drowsiness or lack of sleep does not have the potential for catastrophic consequences, unlike the airline pilot who is responsible for dozens or hundreds of lives. While some may not be entirely satisfied with all aspects of the fatigue rule mandated under HR 5900, I have to say that the overall result is indeed historic. More than 20 years of political gridlock has been broken. What we have today are a set of rules that reflect our modern understanding of fatigue and target operational risks based on science rather than just political horse-trading. The new rule gives the industry the flexibility to respond to new operational risks as they emerge and to embrace new advances in science as appropriate. The world has taken note, and to a great extent is following this positive example.

PILOT CERTIFICATION AND QUALIFICATION REQUIREMENTS FOR AIR CARRIER OPERATIONS

There is still a great deal of work to be done to address the issues of experience and training requirements called for under HR 5900. Congress and family groups were right to call for action in this area. The training standards for airline operations have been relatively static for decades, while incremental implementation of new technology and wholesale restructuring of the industry has caused those training requirements to become dangerously outdated and we are seeing some tragic consequences. This problem is not unique to the United States. It is a worldwide problem that must be dealt with now and the world will be watching to see how the FAA addresses it.

Since the initial proposal of the 1,500 hour rule, the Flight Safety Foundation expressed reservations regarding the focus that HR 5900 places on mandatory flight hours. Our position has been that if a flight crew needs to know, understand or have a specific skill set in order to protect the lives of their passengers, then steps should be taken to ensure the knowledge is obtained through training or previous experience. Mandating an arbitrary number of hours experience required to be in a cockpit makes the dangerous assumption that specific knowledge will be obtained simply due to hours in the air. This leaves too much to chance.

There are countless examples of pilots with many thousands of hours, who lacked the critical knowledge to avert a tragedy. The Air Florida pilots who crashed at Washington National more than 20 years ago had 8,300 hours and 3,500 hours respectively, yet still lacked critical knowledge of cold weather and de-icing operations. While the final report has not been issued, it is clear that there were gaps in the knowledge of the crew of Air France 447, which crashed in the Atlantic Ocean several years ago, that lost control of the aircraft following a brief failure of an airspeed sensor.

While the purpose of a 1,500 hour rule is understood, the Flight Safety Foundation strongly supports the notion that a structured training program can allow this requirement to be reduced, since that training program would reduce risk by leaving less to chance. The Foundation believes the real effectiveness of the new rule will be more a result of mandating critical training that targets risk in the real world, rather than simply increasing the number of hours.

The Notice of Proposed Rule Making (NPRM) calls for several changes that are very important. The requirements for an Airline Transport Pilot (ATP) Certification Training Program and the development of a revised ATP Knowledge are positive. These changes will help ensure that the basic knowledge of air carrier operations are provided and will open the door to a modernization of the existing knowledge test.

The Flight Operations and Qualifications Aviation Rulemaking Committee (ARC) also recommended a requirement for an Advanced Jet Training Course. The Foundation supports this recommendation since it would ensure specific practical training that addresses many of the highest-risk areas, and would meeting many of the requirements spelled out in Section 216 of HR 5900.

The NPRM also includes two new experience requirements that institutionalize common practice and make a great deal of sense. One is the requirement for 50 hours of multi-engine experience for a second in command. This requirement is a de facto industry standard already. The other requirement is for 1,000 hours of airline operational experience before assuming the role of captain. This amounts to a requirement for one to two years of line experience as a first officer before assuming command. Most responsible airlines already have a similar requirement.

It is vitally important to ensure that Captains are promoted in a thoughtful and deliberate manner. Captains become mentors and trainers for first officers.

They also set the limits of what is acceptable and the types of risks that may or may not be taken. The professionalism and maturity of the Captains drive the safety culture of an airline. I hope to see additional work that focuses on the important area of mentoring.

LOOKING FORWARD

The progress in the area of fatigue and pilot training is impressive. However I would like to offer some suggestions and cautions as we move forward with further implementation.

First of all I would like to suggest that while external forces have provided an essential push, it is important at this junction that the experts are allowed to focus on the issues that pose a risk rather than just the issues that resonate politically. For example, HR 5900 Section 207 directs the FAA to, "...conduct a study on aviation industry best practices with regard to flight crewmember pairing, crew resource management techniques, and pilot commuting." I have heard a lot of controversy about pilot commuting but I have heard very little about crew pairing. As a safety professional I will tell you crew pairing techniques are a powerful tool to mitigate risk. We are at a juncture where the things that capture the public's imagination may not be the same things that need to be addressed to avert the next tragedy.

Secondly, progress against the objectives of HR 5900 must be made with a sense of urgency, but without overreaching. The myriad of working groups, studies and deadlines, specified by HR 5900 had the desired effect. They created a flurry of activity that unfroze the system that had and drove overdue action on key issues. However, this flurry of activity generated as much heat as it did light. The breadth of activity clearly exceeded the resources and capability of the FAA to manage. From this point on, it would be best if the sense of urgency could be maintained and a schedule agreed to that allows for the thoughtful implementation of some difficult changes.

Finally, it should be acknowledged that HR 5900, and specifically the requirement for an ATP for the second –in– command in an air carrier, will have significant international implications. Since the close of World War II, the United States has been a leader in the field of aviation, and the FAA has served as a model for regulatory authorities around the world. For the first time, the FAA will promulgate a rule that the rest of the world will have to universally dismiss. Given the structure of the global airline industry and the demand for aviation professionals around the world, it will be impossible for foreign regulators to follow the FAA's lead and implement an ATP requirement for the second-in-command of an air carrier.

In addition, there will be challenges for foreign regulators who are confronted with an Airline Transport Pilot Certificate that was issued with less than 1,500 hours. While the Flight Safety Foundation strongly supports the FAA proposal to issue an ATP Certificate with restricted privileges at 1,000 hours, such a proposal does fall outside the framework of international standards and will require some thoughtful adjustments to accommodate international operations.

These are not inconsequential concerns. Our actions reverberate throughout the world and it is important that our lawmakers and regulators at the FAA consider the impact new regulations have on international operators flying into the US as well as the impact some of these regulations may have on the US's position as the standard-bearer for aviation regulation. We are encouraged by many of the advances we have seen during the implementation of HR 5900, but still see some areas that need more work.

For the traveling public, the tragedy of an aviation accident is magnified by the rarity. We've achieved levels of safety that are the envy of other industries and can make all of us proud. But we must stay vigilant and understand that safety is an ongoing effort. As we get deeper into the implementation of HR 5900, it is our hope that the FAA continues to increase its work with safety professionals and other experts in order to make our safe skies even safer.

Thank you very much for this opportunity to testify this afternoon.

In: Pilot Training and Safety
Editor: David M. Fontenelle

ISBN: 978-1-62417-400-1

Chapter 4

STATEMENT OF CAPTAIN CARL KUWITZKY, PRESIDENT, COALITION OF AIRLINE PILOTS ASSOCIATIONS. HEARING ON "COMMERCIAL AIRLINE SAFETY OVERSIGHT"*

Good afternoon, Madame Chairperson Cantwell, Senator Thune and the distinguished Members of the Subcommittee on Aviation Operations, Safety and Security as well as Commerce Committee Chairman Rockefeller. My name is Captain Carl Kuwitzky. I am a line pilot currently flying for Southwest Airlines for 29 years. I am honored and grateful for your invitation to appear before you today in my capacity as President of the Coalition of Airline Pilots Associations (CAPA).

CAPA is a trade association focused exclusively on the safety of the flying public and the enhancement of the piloting profession. We represent over 31,000 commercial airline pilots of Southwest Airlines, American Airlines, US Airways, UPS, Southern Air, ABX Air, Atlas Air Cargo, Kalitta Air, Polar Air Cargo, Arrow Air, Horizon Air, Miami Air, USA 3000, Omni Air, and Gulfstream Air. CAPA was proudly in the forefront in the fight for the passage of the Airline Safety and Federal FAA Extension Act of 2010.

We congratulate this Subcommittee and all of the Senate and the House for enacting this important legislation vital to the safety of America' flying

* This is an edited, reformatted and augmented version of a Statement Presented March 20, 2012 before the Senate Committee on Commerce, Science, and Transportation, Subcommittee on Aviation Operations, Safety & Security.

public and to maintaining and improving the high stands of our profession. Enactment of the Airline Safety Bill in 2010 was indeed a watershed moment for the commercial aviation industry. However, it was only the first step that needs to be taken for the safety of our flying public. The rigorous implementation of that bill by the Federal Aviation Administration is the next step that must be taken to ensure commercial aviation safety. As you know, the FAA recently released the new flight duty and rest regulations required by the Airline Safety and FAA Extension Act of 2010.

These long awaited regulations were on the National Transportation Safety Board's "Most Wanted" list and sought by professional commercial airline pilots for over twenty years. The FAA correctly used all available science, convened an Aviation Rulemaking Committee and considered thousands of public comments to determine specific rules governing flight time and duty time in our profession.

While CAPA applauds the promulgation of those rules, it is with regret that we must suggest that the critical standard of "One Level of Safety" for all commercial carriers was not met by the FAA. The regulations unfortunately have a critical flaw, that is, the exclusion of mandatory compliance by "all-cargo" operations. The safety of our nation's air space is only as strong as its weakest link, and that weak link is fatigue in primarily all-night cargo operators who were excluded from the rule.

Madame Chairperson, approximately 15% of all departures in the United States are all-cargo flights. These aircraft fly in the same airspace and routinely interact with passengers carriers throughout the aviation system. Cargo and passenger aircraft interact during numerous critical phases of flight which include Precision Radar Monitored (PRM) approaches and Land and Hold Short Operations (LASHO). During PRM approaches, aircraft fly with absolute minimum separation, relying on each other to fly precise approaches. During LASHO operations, aircraft are cleared to land and hold short of a crossing runway or taxiway where other aircraft are operating.

Madame Chairperson, an exemption similar to this cargo "carve-out" was attempted in the early 1990's when TCAS was mandated for passenger carriers while cargo carriers were exempt due to the alleged cost of the system. Following a fatal passenger/cargo midair accident in 1996, a near miss in 1997 between a UPS Boeing 747 and Air Force One, and two additional near misses in 1999, TCAS was finally mandated for all cargo aircraft. We feverently hope that the current cargo carve-out can be obviated by the FAA or by Congressional statute if necessary before there is a terrible episode of a fatal accident. It is CAPA's position that until the new flight and rest regulations are applied to passenger carriers, all-cargo carriers and supplemental carriers, our air space will not be governed under "One Level of Safety" and the American public will not be assured of their right to a maximum level of safety.

The Airline Safety and Federal FAA Extension Act of 2010 was also enacted to upgrade the minimum professional stands for First Officers piloting commercial aircraft. CAPA has always advocated the Airline Transport Pilot (ATP) License as the minimum standard for employment as a pilot with a Part 121 or Part 135 air carrier. The ATP provides the minimum requisite academic coursework, flight training and experience needed for the safe piloting of today's complex, high-speed aircraft through a congested, multifaceted air traffic control network in difficult weather and other situational environments. Mainline air carriers require the ATP for employment and once again, "One Level of Safety" dictates that all air carriers, regional or otherwise, should require the ATP as well. The 1500 flight hours that the ATP requires develops a mature, experienced and professional aviator who has the foundation to exercise prudent judgment while responsible for the safe transportation of scores of passengers.

We were gratified that the Airline Safety bill did suggest directionally that all commercial airline pilots have an ATP. We were concerned that the legislation did allow the FAA Administrator to make exemptions from this standard. We were further concerned that the FAA rules will, while upgrading the current minimum standard, not actually require all commercial aviation pilots to have an ATP. We will continue to press the FAA and the Congress to ultimately fix this deficiency in the standard required in our profession.

Madame Chairperson, on another issue not part of the Airline Safety Bill, I would like to raise the issue of the Federal Flight Deck Officer program. The FFDO program is a highly cost effective one designed to provide the last line of defense against possible terrorist activity aboard commercial aircraft. The pilot participants in this program bear significant costs out of their own pockets to train and take part in the program. Other programs such as that of the Federal Air Marshalls (FAMS) are hugely augmented, again in a highly cost effective manner, by the FFDO pilots. The Federal expenditures associated with the FFDO program are a tiny fraction of the cost of the FAM program. As you know, the FFDO budget has been stagnate for a number of years resulting in a significant backlog of pilots who have applied to participate in the program and are awaiting training due to the lack of federal funds. In addition, unfortunately the budget the President submitted recently to the Congress cuts very substantially the federal funds allocated to the FFDO program.

CAPA urges the Senate Commerce Committee and the Senate Appropriations committee to reject that cut in expenditure and indeed increase FFDO funding so new applicants can be trained for the program.

Finally, CAPA congratulates this Subcommittee, the full Commerce Committee and the entire Congress for finally enacting an FAA Reauthorization bill. This bill will help to propel NextGen forward and provide a sustainable, certain level of funding for the nation's air travel infrastructure. It is the first step the Congress has taken to the development of a national air transportation policy for the 21st century.

CAPA looks forward to working constructively with this Subcommittee on all of the matters I have raised in my testimony. I look forward to your questions and comments. Madame Chairperson, I would respectfully request that my longer statement be included in the record of this hearing.

INDEX

D

E

P

Q

R

S

T

U

V

W